Satellite-Based Earth Observation

Christian Brünner · Georg Königsberger
Hannes Mayer · Anita Rinner
Editors

Satellite-Based Earth Observation

Trends and Challenges for Economy and Society

Springer

Editors
Christian Brünner
Competence Centre Space Law/Space Policy, Subpoint NPOC Austria/ECSL
University of Graz
Graz, Steiermark
Austria

Hannes Mayer
Competence Centre Space Law/Space Policy
University of Graz
Graz, Steiermark
Austria

Georg Königsberger
Chamber of Commerce Steiermark
Graz, Steiermark
Austria

Anita Rinner
Competence Centre Space Law/Space Policy
University of Graz
Graz, Steiermark
Austria

ISBN 978-3-030-9079-1 ISBN 978-3-319-74805-4 (eBook)
https://doi.org/10.1007/978-3-319-74805-4

Printed on acid-free paper

This Springer imprint is published by the registered company Springer Nature Switzerland AG
The registered company address is: Gewerbestrasse 11, 6330 Cham, Switzerland

Foreword I

When distinguished space exploration experts meet for a major symposium at the Styrian Economic Chamber, they have good reason to do so. After all, as the 'company of entrepreneurs', it is the ideal venue for an exchange between the scientific and business communities. After all, satellite-based earth observation is becoming increasingly significant in many areas of business and daily life. From navigation and weather forecasting to investigation of soil properties and the satellite data needed for autonomous driving, working with space data is essential in many fields of business and research. Promoting the transfer of technology from space science to industrial fields and thus to new applications contributes significantly towards a region's attractiveness—and this is especially true for Styria. An R&D quota of 5, 16%, one of the highest in Europe, is testimony to the fact that research and development play a prominent role in Styria. The mobility cluster, which includes the major area of space technology, has an R&D quota of almost 12%. And the significance for our companies of research conducted at Styrian institutions of higher education, is illustrated by another, equally impressive, figure: 186 million euros in third-party funds has recently been poured into Styrian universities and institutions of higher education. This is money invested in the region by companies in order to achieve success nationally and internationally through innovation. And, in doing so, these companies are clearly stating their commitment to Styria as a business location, by creating and maintaining jobs and thereby guaranteeing the region's prosperity.

Steiermark, Austria

Josef Herk
President, WKO Steiermark
josef.herk@wkstmk.at

Josef Herk has been the President of the Styrian Economic Chamber since 2011, and, in this capacity, represents about 74,000 companies in the Austrian province of Styria.

He also runs a medium-sized automobile body and paint shop, a family business now in its second generation.

Foreword II

In 2017, we celebrated the 60th anniversary of the launch of *Sputnik*. *Sputnik* was the first human-made object in orbit. Shortly afterwards, when the first cameras were put on satellites, earth observation from space was born. The concept of giving eyes to satellites was mainly inspired by the dramatic 'space race' between the USSR and the US during the Cold War. In the beginning of the 1970s, the transition from purely military use of earth observation technology to the civilian domain was marked by the science-driven *Landsat* program. *Landsat* served the emerging demand for comprehensive information on the planet's environmental health. Although earth observation, or at least a significant part of it, had now left the realm of defence, it was still the domain of an exclusive circle of scientists and 'nerds'.

2017 also saw the 40th anniversary of the launch of *Meteosat*. *Meteosat* was the first fully operational earth observation satellite built as a joint European project. A new era of weather and climate forecast was dawning and brought observational data about the earth into people's minds and living rooms. A typical weather forecast at the end of the news now included animations tracing the formation and movements of clouds in the diurnal cycle, featuring photos of continents rather than schematic graphics. *Meteosat* was followed by the first European Radar Satellites ERS-1 and -2 and ENVISAT, all three truly European endeavours. These steps culminated in the European earth observation programme *Copernicus,* which has been operational since 2014. *Copernicus* provides a comprehensive system of space infrastructure and offers sustainable services and information products for monitoring the environment and supporting civil security. With *Copernicus*, earth observation crossed the threshold to benefiting the public, evolving into an indispensable tool for tackling grand societal challenges.

At the same time, the entire space sector is facing a giant leap in evolution both in economic and institutional terms, often referred to as *New Space* or *Space 4.0*. More and more stakeholders are active in space, the private and commercial sectors are ramping up and space is increasingly becoming a 'normal' business, like the automotive sector, for example. The earth observation sector is more attractive than ever. In the coming years thousands of—predominantly small—satellites will be launched, based on digital technologies and facilitating new digital business models.

This is why the book 'Trends and Challenges of Satellite-based Earth Observation for Economics and Society' comes exactly at the right time. The editors *Christian Brünner, Georg Königsberger, Hannes Mayer* and *Anita Rinner* and all the contributors succeed in addressing the right questions and discussing appropriate issues. Earth observation systems such as *Copernicus* were designed and created by scientists and engineers. A particular asset of this book is seeing earth observation from a multitude of perspectives, not only from the technical side.

Wien, Austria

Thomas Geist
FFG—Austrian Research Promotion Agency
Austrian delegate to the ESA Programme
Board for Earth Observation and the Copernicus Committee
thomas.geist@ff.at

Thomas Geist started working for the Aeronautics and Space Agency (ALR) within the Austrian Research Promotion Agency (FFG) in 2007.

As an expert for Earth Observation (EO) Thomas is the Copernicus contact point in the agency and responsible for the coordination of EO projects in the national R&D funding programme ASAP (Austrian Space Applications Programme). Thomas represents Austria in the ESA Programme Board for Earth Observation (PB-EO) and since 2017 he is the Chairman of the Data and Operations Scientifical and Technical Advisory Group (DOSTAG) at ESA. Furthermore, he is delegate in the Horizon 2020 Space Programme Committee and the Copernicus Committee.

Before joining FFG, Thomas worked as a senior researcher at the University of Innsbruck. He holds a Master's degree in Physical Geography from the University of Munich, and a Ph.D. degree from the University of Innsbruck (Faculty of Geo- and Atmospheric Sciences).

Preface

Outer Space is not only fascinating but—since the first launch of a satellite, the Russian Sputnik in 1957—has become an ever more attracting target and object of activities. Apart from the exploration of outer space, the utilization of outer space for different purposes has been in the centre of interest from the very beginning. In fact, 'outer space' has meanwhile turned into one of the most important—and most indispensable—fields of human activities, of society and of politics. The main applications are communication, navigation and Earth observation, furthermore in the near future probably the exploitation of natural resources ('space mining').

Nowadays the results of space science, space activities and space technology are applied in countless areas of everyday life to such a degree that mankind is depending on them. Each of us uses at least 20 different satellite applications everyday, this especially in the fields of weather forecast, navigation and communication; less visible—but likewise important—are, for e.g. applications for steering systems in transformer stations for electricity.

Furthermore, the importance of space applications for bridging the digital divide in order to enhance economic and societal development throughout the world has to be mentioned. 54% of the world population does not have access to Internet; the respective figures for Africa and for Europe are 75%(!) and 21%.

And yet, the strategic as well as operational importance of outer space has not reached its top, it is incessantly growing. Keywords—especially in the context of commercialization of outer space—are 'New Space', 'Astropreneurship' or 'Space 4.0', connected with new models of Public–Private Partnership. Parallel to this development, there is a growing need for an effective political, societal, legal, but also ethical framework, this on an international as well as on a national level.

An important task in this context is the education, information and participation of a broad public in order to raise awareness of the importance of exploration and utilization of outer space ('Outreach').

Numerous organizations and institutions are active for and in outer space. Besides governmental ones—for e.g. the United Nations with several space-related programs and the Committee on the Peaceful Uses of Outer Space (UNCOPUOS) with its two Subcommittees (the Legal one and the Scientific and Technical one);

the European Space Agency (ESA); (National) Space Agencies—nongovernmental organizations, private companies, experts, etc., play an important part in order to shape the above mentioned framework.

Among the nongovernmental organizations are the International Astronautical Federation (IAF), the International Academy of Astronautics (IAA) and the International Institute of Space Law (IISL). They are the organizers/co-organizers of the annual International Astronautical Congress (IAC). Furthermore, the European Centre for Space Law (ECSL) has to be mentioned. I have the privilege to be a member of IAA, IISL and ECSL.

The ECSL was founded 1989 by Gabriel Lafferanderie, a pioneer of space law, within the organizational frame of ESA. Among the most important activities of ECSL count the annual Summer Course on Space Law and Space Policy, which is usually held in different countries and in cooperation with one of the ECSL National Points of Contact (NPOC), as well as the regular organization of the Practitioner's Forum and recently the Young Lawyer Symposium (YLS).

On a national level, the work of ECSL is supported by the National Points Of Contact (NPOC). In 2001, I established the Austrian NPOC at the University of Graz which was possible due to the continuing affords of the ECSL's former chairman Gabriel Lafferanderie and the support from Eva-Maria Schmitzer, at the time Head of the Space Research Division at the Federal Ministry for Transport, Innovation and Technology. An important contribution to the implementation of this idea came also from Leopold Summerer, who had presented a report on the dissemination of Austrian expertise in the field of space law in 2001.

I founded NPOC Subpoints at the Austrian universities Linz, Salzburg, Innsbruck, Wien, at the Danube University Krems and the Vienna University of Economics and Business. I headed the NPOC Austria until 2009, when I handed it over to the University of Vienna. Since then the University of Graz is a Subpoint of ECSL/NPOC Austria, which we call 'Competence Centre for Space Law and Space Policy (CC)'. The Subpoint, respectively, the CC do not have an organizational basis, but a personal one, including Georg Königsberger, Hannes Mayer, Anita Rinner and Katharina Zollner.

By the way, my curiosity about matters of outer space was wakened in 1991. As a member of the Austrian Parliament and as the spokesman for academic affairs of the Austrian People's Party (ÖVP), I had the possibility to watch the launch of the Sojus—Rocket from the Cosmodrome in Baikonur, Kazakhstan, with the Austrian astronaut Franz Viehböck on board.

Since its founding, the NPOC Austria and/or the Subpoint Graz, respectively, made it its business to promote teaching and research in the field of space law and space policy. A Seminar on Space Law and Space Policy is held annually within the diploma program of law. Master's and doctoral theses are supervised.

Moreover, every year since its founding, the NPOC could offer several students of Austrian universities the possibility to take part in the ECSL Summer Course on Space Law and Policy as well as in other space-related events. Due to an agreement with the former head of the Bureau of International Law within the former Federal Ministry for Foreign Affairs, Ambassador Dr. Hans Winkler, students can participate as part of the Austrian delegation to the meetings of the Legal Subcommittee of UNCOPOUS.

Furthermore, it is an important task to circulate information on space (law) topics and events among a broad public. Awareness of the importance of space exploration and space utilization for everyday life has to be raised.

My approach is an interdisciplinary one. Not only legal and political perspectives are reflected, but also societal, cultural, economic, technical and ethical ones.

Several symposia have been organized since 2005. The recent symposium 'Trends and Challenges of Satellite Based Earth Observation for Economy and Society' was organized by the NPOC Subpoint Graz, respectively, by the Competence Centre for Space Law and Space Policy.

The symposium consisted of two parts: The presentations by lecturers and moderators and their professional contributions to the topic 'Satellite-Based Earth Observation' (see the program under the link https://homepage.uni-graz.at/de/christian.bruenner/. Remark: Not all lecturers were able to deliver the written version of their papers. Fortunately Silke Migdall, Lena Brüggemann and Heike Bach, and Michael Wurm filled in the gap). And: 'Space Art & Space Technology', an exhibition of paintings of selected artists and of a model of the nanosatellite OPS-SAT took place, furthermore the video 'ART_SAT' was presented (see the contribution of Richard Kriesche and my introduction page xxi).

Important support came from the Styria Economic Chamber (WKO/Steiermark) and its President Ing. Josef Herk. Furthermore, the symposium, respectively, the publication was sponsored by Austrospace, the FFG/NPOC Austria, the BMVIT, the University of Graz, the Government of Styria, the City of Graz and the Bankhaus Krentschker. The lecturers and the moderators contributed substantially to the success of the symposium.

Help came also from the students of the 2017 Seminar on Space Law and Space Policy (see the list of the students https://homepage.uni-graz.at/de/christian.bruenner/). Furthermore, the members of the NPOC Subpoint Graz, respectively, the Competence Centre did a great job.

All actors and sponsors deserve my sincere thanks.

Last but not least, I want to thank the publishing house Springer Wien New York and its representative, Ms. Silvia Schilgerius, Senior Editor Applied Sciences. The kindness and cooperation have been outstanding.

Graz, Austria Christian Brünner

WKO
STEIERMARK

FFG
Austrian
Research Promotion Agency

bmvit
Federal Ministry
for Transport,
Innovation and Technology

ECSL
European Centre for Space Law
National Point of Contact Austria
NPOC Austria

UNI
GRAZ

BANKHAUS
KRENTSCHKER

Contents

Editors and Contributors

About the Editors

Christian Brünner is Professor Emeritus of Public Law at the University of Graz. He is also a Lecturer of Space Law and Space Policy at the University of Graz, the ECSL Summer Courses and the ISU Space Studies Program.

He was Rector of the University of Graz, Chairman of the Austrian Rector's Conference, Member of the Austrian National Parliament and of the Styrian Regional Parliament.

He is founder of the Austrian National Point of Contact (NPOC), ECSL/ESA, which he chaired for several years. Furthermore, he is member of the ECSL Board, the IISL, the International Academy of Astronautics and the Austrian Delegation to UNCOPUOS.

Besides publications in the fields of education(al law), university management, control, parliamentarianism and religious and ethnical minorities, he has authored and/or edited several publications in space law, including the book (as Co-Editor) Outer Space in Society, Politics and Law which was awarded the Social Sciences Book Award of the International Academy of Astronautics 2012. Furthermore, he was awarded the Polarstern Award 2015 by the Austrian Space Forum (Österreichisches Weltraum Forum).

He is active in various fields to raise public awareness of space, space law and space policy.

Georg Königsberger is Legal Officer at the Styrian Economic Chamber's labour and social law desk. Among his earlier career steps was a stint as scientific project contributor at the Institute of Austrian, European and Comparative Public Law, Political Science and Administrative Studies at Karl-Franzens-University Graz. He is a contributor of the Competence Centre for Space Law and Space Policy at Karl-Franzens-University Graz. He studied law and history at Karl-Franzens-University Graz and Charles University Prague. Publications in the fields of labour law; teacher in adult vocational training.

Hannes Mayer works at the Competence Centre for Space Law and Space Policy as well as the Institute of Canon Law of Karl-Franzens-University Graz. He serves as a member of the Austrian delegation to the United Nations Committee on the Peaceful Uses of Outer Space (UNCOPUOS). He is a member of the European Centre for Space Law, the Space Generation Advisory Council's Space Law and Policy Project Group, the Austrian Space Forum and the History Committee of the International Academy of Astronautics (IAA). He is a board member of the Diocesan Committee of Catholic Organizations Styria and a member of the European Academy of Religion. He has (co-)authored several publications on outer space and is a member of the editorial staff of the Theologicum Magazine.

Anita Rinner is a Trainee Lawyer (with bar examination) at Muhri & Werschitz Attorneys, Graz; Lecturer in Space Law and Policy, University of Graz; Member, Competence Centre for Space Policy and Space Law, Graz; Member, European Centre for Space Law (ECSL); Member, Austrian Space Forum, and Member, International Institute of Space Law (IISL). Austrian Delegate, United Nations Committee on the Peaceful Uses of Outer Space Legal Subcommittee, Vienna; Alumna, European Centre for Space Law Summer Course (ECSL), Rijeka; Tutor, ECSL Pafos; Alumna, International Space University Space Studies Program, Strasbourg; Oral and written Presentations at the International Astronautical Congress in Beijing, Toronto and Jerusalem; Presentations at various space-related symposia in Graz, Vienna, Berlin and Rom; Participant of ESA/CNES Parabolic Flight Campaign, Bordeaux.

Contributors

Heike Bach VISTA Remote Sensing in Geosciences GmbH, Munich, Germany

Ulrike Bohlmann ESA, Paris, France

Peter Breger DG for Internal Market, Industry, Entrepreneurship and SMEs, European Commission, Brussels, Belgium

Lena Brüggemann VISTA Remote Sensing in Geosciences GmbH, Munich, Germany

Christian Brünner University of Graz, Graz, Austria; European Centre for Space Law/ESA, Paris, France

Lorenz Brunner Federal Ministry for Europe, Integration and Foreign Affairs, Vienna, Austria

Holger Flühr Institute of Aviation, FH JOANNEUM (University of Applied Sciences), Graz, Austria

Norbert Frischauf CERN, Geneva, Switzerland; Austrian Space Forum (OEWF), Vienna, Austria

Klaus Granica Research Group Remote Sensing and Geoinformation, Forschungsgesellschaft Joanneum MbH, Graz, Austria

Werner Hauser FH Joanneum, Graz, Austria; Alpen-Adria-Universität Klagenfurt, Klagenfurt, Austria

Manuela Hirschmugl Research Group Remote Sensing and Geoinformation, Forschungsgesellschaft Joanneum MbH, Graz, Austria

Christian Hoffman GeoVille GmbH, Innsbruck, Austria

Rudolf L. Hütter Amt der Steiermärkischen Landesregierung, A17 Landes- und Regionalentwicklung, Referat Statistik und Geoinformation, Graz, Austria

Yvonne Karimi-Schmidt Institute of International Law and International Relations, Karl-Franzens-University Graz, Graz, Austria; European Centre for Space Law, Paris, France

Richard Kriesche Kulturdata-Graz, Graz, Austria

Georg Königsberger Legal Department, Styrian Economic Chamber, Graz, Austria

Maria Lemper GeoVille GmbH, Innsbruck, Austria

Hannes Mayer Competence Centre for Space Law and Space Policy, Institute of Canon Law, Karl-Franzens-University Graz, Graz, Austria

Silke Migdall VISTA Remote Sensing in Geosciences GmbH, Munich, Germany

Martin Mössler ESA BIC, Graz, Austria

Thomas Neger Neger/Ulm Rechtsanwälte GmbH, Graz, Austria

Gina Petrovici University of London, London, UK

Wolfgang Rathgeber European Space Agency (ESA), Frascati (RM), Italy

Anita Rinner Rechtsanwaltsanwärterin, Muhri und Werschitz Partnerschaft von Rechtsanwälten GmbH, Graz, Austria; Lehrbeauftragte für Raumfahrtrecht und Raumfahrtpolitik, Karl-Franzens-Universität, Graz, Austria; Muhri & Werschitz Attorneys, Graz, Austria; Karl-Franzens-University, Graz, Austria

Mathias Schardt Research Group Remote Sensing and Geoinformation, Forschungsgesellschaft Joanneum MbH, Graz, Austria; Graz University of Technology, Institute of Geodesy, Graz, Austria

Wolfgang Schinagl WKO Stmk, Technische Infrastruktur, Digital Content Research & Development Center (DCRDC), Graz, Austria

Gerhard Schnedl Institute of Public Law and Political Science, Karl-Franzens-University Graz, Graz, Austria

Lesley Jane Smith Leuphana University, Lüneburg, Germany

Alexander Soucek ESA, Paris, France

Wolfgang Sulzer Department of Geography and Regional Sciences, University of Graz, Graz, Austria

Alessandra Vernile Eurisy, Paris, France

Gudrun Waniek Schönherr Rechtsanwälte GmbH, Vienna, Austria

Michael Wurm German Aerospace Center (DLR), Earth Observation Center (EOC), Oberpfaffenhofen, Germany

Katharina Zollner (Editorial assistance to the book in hand) Competence Centre Space, Law/Space Policy, University of Graz, Graz, Steiermark, Austria

Introduction

For as long as humankind has existed, men and women have looked at the sky. With respect and highly fascinated, they have watched the rising and setting of the sun and of the moon and the stars glittering in the dark sky.

Looking at the sky, their thoughts have circled questions like: Where do the celestial bodies come from? Is there an entity—God or whatever names were given to the unknown—which has created all this and has implemented the rhythm which we can experience in our lives and nature? Or: Do the sun, the moon and the stars influence our lives? Or: Who am I in this immense and unknown universe?

In the ancient Egyptian, Greek and Roman mythologies the celestial bodies were seen as Gods. Astrology, which is 5000 years old, teaches that the planets including the sun and the moon are symbols representing certain aspects of human's personality. For Hermes Trismegistos, a philosopher in ancient Egypt, it was a 'natural law' that how it is aloft, so it is beneath.

The celestial bodies, especially the sun and the moon, have been early objects of astronomical calculation. For millennia, astronomers, astrologers and philosophers have been developing theories attempting to explain the material world, based upon astronomical calculations. They were also used for practical terrestrial needs, for instance for navigation on the seas.

The early space-oriented activities already show the two main views, when we talk about space: on the one hand, the view from Earth 'outside' with the objective of exploring the universe, and, on the other hand, the view towards Earth to observe the place where we live, an observation that serves manifold purposes.

Besides navigation, telecommunication and the recently planned exploitation of natural resources on the moon and asteroids ('space mining') earth observation is the largest sector of space applications. Around 30% of the whole budget of ESA, the European Space Agency, is spent for Earth observation activities. Furthermore, the objectives of earth observation are multifold. In a list of space applications meeting societal needs ESPI, the European Space Policy Institute, mentions several objectives, for instance food management, land management, urban planning, preserving archeological sites, forest management, monitoring of freshwater, assessing the risks of natural disasters, search and rescue, national security, etc.

An important aspect of the framework of earth observation is how the access to gained and generated data is regulated. For example, the data which are gained by the EU-Project Copernicus are freely accessible and without costs.

Earth observation's genius loci is Wilhelm Nordberg. He was born in Fehring and has studied physics at the University of Graz. After his studies, he emigrated to the USA, where he served as NASA's Director for Space Applications for 2 years. In 1955, he was the first to publish his conviction that weather forecast via satellites must be possible. This hypothesis was realized by the weather-satellite TIROS I. Consequently, he turned towards earth observation. A result was the satellite LANDSAT 1. Due to these activities, it is appropriate to consider Willi Nordberg the father of earth observation. The symposium was honoured by the presence of Kurt Nordberg, Willi Nordberg's brother.

With the symposium, we pursued three objectives:

1. We want to give information to a broad public concerning different applications of earth observation data in economy and society and concerning its political, economic and legal framework, this also on the level of the EU (Part I-V).
2. The Outer Space Treaty of 1967 lays down the purpose of space activities. According to Article I par 1, the exploration and use of outer space, including the Moon and the celestial bodies, shall be carried out for the benefit and in the interests of all countries, irrespective of their degree of economic or scientific development, and shall be the province of all mankind.
   The provision is the starting point for a multidimensional view of space activities. In 'Space Art & Space Technology', an exhibition of paintings and installations by Edith Temmel, Petra dieHolasek, Hannes Scheucher, Josef Bramer und Günther Friesinger, the nanosatellite OPS-SAT, respectively, TUGSAT-1/BRITE-Austria and the presentation of the video clip ART_SAT we outline the cultural/artistic and technical dimension (Part VII).
   A few words concerning ART_SAT (for more information see the contribution of Richard Kriesche in this volume): ART_SAT was—besides 15 scientific experiments—an artistic experiment within the project AUSTROMIR, the mission of an Austrian astronaut to the Russian Space Station MIR. It was carried out in cooperation between Richard Kriesche, the designer and operator in the ground station, and Franz Viehböck, the Austrian astronaut on board of MIR.
3. As university teacher, it is important for me to fill young people and students with enthusiasm for outer space in young people and students and to instruct them properly. Anita Rinner, space law lecturer in the Seminar on Space Law and Space Policy which is held annually at the University of Graz, proposed to invite the students of the 2017 seminar to draw up posters and present them in a poster session during the symposium (see Part VI).

Part VIII of this volume is dedicated to my 75th birthday. The part is based upon a proposal of my co-workers at the Subpoint Graz/Competence Centre for Space Law and Space Policy. Initially I hesitated, but then I agreed, and now I am grateful for this Part.

Christian Brünner

Part I
Earth Observation and Its Multidimensional Application

The European Space Agency's Earth Observation Programme

Wolfgang Rathgeber

1 Introduction

The European Space Agency (ESA) has been managing an Earth Observation Programme since the launch of its first Meteosat meteorological satellite in 1977. Following the success of this first mission, a subsequent series of Meteosat satellites, two Earth Remote Sensing satellites (ERS-1, ERS-2), and the Envisat mission provided a wealth of valuable data about Earth, its climate and changing environment. As a result, ESA has become a major provider of Earth observation data and stimulated the build-up of the Earth science community and beyond.

Started in the mid-1990s, the ESA's Living Planet Program heralded a new approach to satellite observations for Earth science with focused missions defined, developed, and operated in close cooperation with the worldwide scientific community. As time progressed, three main categories of ESA Earth observing missions have emerged: Earth Explorer, Earth Watch, and Copernicus (formerly GMES) Sentinel missions.

ESA also distributes satellite data from international partner agencies, so called Third Party Missions (TPM). The data from these missions are distributed under specific agreements with the owners or operators of those missions, which can be either public or private entities outside or within Europe.

W. Rathgeber (✉)
European Space Agency (ESA), Largo G. Galilei 1, 00044 Frascati (RM), Italy
e-mail: wolfgang.rathgeber@esa.int

C. Brünner et al. (eds.), *Satellite-Based Earth Observation*,
https://doi.org/10.1007/978-3-319-74805-4_1

2 ERS and Envisat

Both these missions addressed a wide range of environmental questions and predate the formal establishment of the Earth Explorer and Sentinel missions. The first European Remote Sensing Satellite (ERS-1) was launched in 1991. It was ESA's first sun-synchronous polar-orbiting remote sensing mission. ERS-1 carried a comprehensive payload including an imaging Synthetic Aperture Radar (SAR), radar altimeter, and other instruments to measure ocean surface temperature and winds. In March 2000, nine years after launch, a computer and gyro control failure led to ERS-1 finally ending its operations after far exceeding its planned lifetime.

ERS-2, which overlapped with ERS-1, was launched in 1995 with an additional sensor for atmospheric ozone research. Both satellites collected a wealth of valuable data on Earth's land surfaces, oceans, and polar regions that were used to monitor natural disasters such as severe flooding and earthquakes in remote areas of the world. Shortly after the launch of ERS-2 in 1995, ESA operated the two satellites in the first 'tandem' mission mode, which lasted for nine months. Beyond the sheer increase in revisiting the same site, this tandem operations boosted SAR-interferometry applications particularly for the generation of digital terrain models (DTMs). In July 2011, ERS-2 was retired and the process of de-orbiting the satellite began.

The giant Environmental Satellite (Envisat) was launched in 2002. The largest Earth observing satellite ever built by any space agency, it was an advanced polar-orbiting satellite that provided measurements of the atmosphere, ocean, land, and ice for over 10 years, double its initially planned lifetime. Envisat's ambitious and innovative payload provided new Earth science data and also ensured the continuity of the data from the two ERS satellites. Exceeding its predicted lifetime by a factor of 3, the end still was sudden. In April 2012, contact with Envisat was lost without warning, likely as a consequence of a failure in the power-regulator or a short-circuit. However, ten years of Envisat's archived data continue to be exploited for studying our planet with demand remaining very high even today.

3 Earth Explorer Missions

The Earth Explorers are research missions designed to address key scientific challenges identified by the science community, while demonstrating breakthrough technology in observing techniques. These missions involve the science community right from the beginning in the definition of new missions followed by a peer-reviewed selection process ensuring that the resulting mission is developed efficiently, and provides the data required by the user. This approach also gives Europe an excellent opportunity for international cooperation with the global scientific community and in the area of technological development. It should be added that some Earth Explorer missions also feature an operational perspective.

GOCE: Launched in 2009, the Gravity field and steady-state Ocean Circulation Explorer (GOCE) mission was the first of the Earth Explorers. GOCE was designed to provide information for understanding critical Earth system variables, such as ocean circulation, sea-level change and variations in ice cover driven by the Earth's gravitational field. The gravitational gradients were measured by a set of six three-axis accelerometers. GOCE provided data to accurately determine global and regional models of Earth's gravity field and geoid, the shape that the surface of the oceans would take under the influence of Earth's gravitation and rotation alone. This advanced research in areas of ocean circulation and ocean dynamics, physics of Earth's interior, geodesy and surveying, and sea-level change.

All mission requirements had been met by the end of 2012. GOCE went on to map gravity signals significantly beyond its original goal of 100 km altitude resolution. The advance meant that for the first time, global ocean currents could be extracted directly from satellite altimetry data. At the end of January 2012, GOCE's orbit was lowered to 235 km, increasing the accuracy and resolution of the measurements to 80 km and improving GOCE's view of smaller ocean dynamics. The satellite was further lowered to 225 km at the end of May 2013 and re-entered Earth's atmosphere in November 2013.

SMOS: Also launched in 2009, the Soil Moisture and Ocean Salinity (SMOS) mission is making global observations of soil moisture and ocean salinity using a unique L-Band Radiometer Microwave Imaging Radiometer with Aperture Synthesis (MIRAS). Soil moisture data are required for hydrological studies and ocean salinity data are vital for improving our understanding of ocean circulation patterns. SMOS completed its nominal three-year mission lifetime in November 2012, but continues to provide valuable results to this day, also beyond its initial mission objectives measuring ice thickness and wind speed.

CryoSat-2: In 2010, Europe's first ice mission CryoSat-2 was launched. Its principal instrument is the Synthetic Aperture Radar/Interferometric Radar Altimeter is designed to measure centimetre-scale changes in thickness of ice floating in the oceans and the ice sheet thickness that blanket Greenland and Antarctica. When combined with other satellite data, analyses showing how the volume of Earth's ice is changing leads to a better understanding of the relationship between ice and the Earth's climate. Cryosat-2 is providing accurate, synoptic, Arctic measurements of ice thickness and column in unprecedented detail.

Swarm: Launched in 2013, Swarm is a constellation of three satellites that provide high-precision and high-resolution measurements of the strength and direction of the Earth's magnetic field using an advanced magnetometer, an accelerometer and an electric field instrument. The geomagnetic field models resulting from the Swarm mission enable new insights into the Earth's interior. They further our understanding of atmospheric processes related to climate and weather, and have practical applications in many different areas such as space weather and radiation hazards.

Aeolus: With a target launch date in 2018, the Atmospheric Dynamic Mission Aeolus will advance global wind profile observations and provide much-needed information to improve weather forecasting. Aeolus will orbit in a Sun-synchronous, dusk/dawn orbit at 408 km and employ a highly sophisticated Doppler wind lidar with a large telescope that collects light backscattered from gas, dust, and droplets of water in the atmosphere. Data from Aeolus are expected to pave the way for future operational meteorological satellites dedicated to measuring Earth's wind fields.

EarthCARE: With a target launch date in 2019, the Earth Clouds Aerosols and Radiation Explorer (EarthCARE) mission is being implemented in cooperation with the Japanese Aerospace Exploration Agency (JAXA). Its aim is to improve the representation and understanding of Earth's radiative balance in climate and numerical weather forecast models. This will be achieved by global measurements of the vertical structure and horizontal distribution of cloud and aerosol fields together with outgoing radiation. The payload comprises two active instruments: a high-resolution atmospheric lidar and radar, and two passive instruments: a multispectral imager and a broadband radiometer. EarthCARE will orbit in an early afternoon sun-synchronous orbit at 393 km.

Biomass: This mission will employ a novel P-band synthetic aperture polarimetric radar operating at 435 and 6 MHz bandwidth. The satellite flies at 637–666 km in a near polar sun-synchronous orbit, and will address one of the most fundamental questions in our understanding of the land component in the Earth system, namely the status and the dynamics of forests, as represented by the distribution of biomass and how it is changing. Biomass will greatly improve our knowledge about the terrestrial carbon pool and fluxes. In addition, the mission responds to the pressing need for biomass observations in support of global treaties such as the United Nations Framework Convention on Climate Change initiative for the Reduction of Emissions due to Deforestation and Forest Degradation.

FLEX: In November 2015 ESA Member States selected the Fluorescence Explorer (FLEX) as the eighth Earth Explorer, following a User Consultation Meeting in September 2015. FLEX is a three-year mission slated for launch in 2022. It will globally monitor the steady-state chlorophyll fluorescence in terrestrial vegetation. FLEX includes three instruments to measure the interrelated features of fluorescence, hyperspectral reflectance, and canopy temperature. FLEX will orbit in tandem with one of the Copernicus Sentinel-3 satellites.

Earth Explorers 9 and 10: For Earth Explorer 9, a call for proposals was published in 2016. Out of the 13 concepts received, two candidates were selected to enter a competitive feasibility phase. These two candidates will spend the next two years being studied thoroughly. In 2019, a User Consultation Meeting will be held, after which a decision will be taken by ESA's Member States as to which of the two contenders will be implemented. Earth Explorer 9 is foreseen to be launched in 2025. A Call for Mission Ideas, initiating the process for Earth Explorer 10, was released in September 2017. The launch of Earth Explorer 10 is foreseen for 2027/28.

4 Earth Watch Missions

The Earth Watch missions are designed to provide Earth observation data for services, including those of the European Organisation for the Exploitation of Meteorological Satellites (EUMETSAT), whose main purpose is to deliver operational weather and climate-related satellite data, images, and products to the European National Meteorological Services as well as to other users worldwide. The ESA-EUMETSAT partnership was established in 1986 and was implemented with a series of geostationary (Meteosat) and polar orbiting (MetOp) satellites, where ESA builds the satellites that are then operated by EUMETSAT.

Meteosat Second Generation (MSG) is a significantly enhanced, follow-on system to the first generation of Meteosat (MFG). MSG consists of a series of four geostationary meteorological satellites, along with a ground-based infrastructure, that will operate consecutively until 2020. The last MSG satellite, MSG 4, was launched in July 2015. The MSG satellites carry two instruments. The Spinning Enhanced Visible and InfraRed Imager (SEVIRI), which has the capacity to observe the Earth in 12 spectral channels and provide image data that are core to operational forecasting needs, and the Geostationary Earth Radiation Budget (GERB) instrument supports climate studies.

The Meteosat Third Generation (MTG) satellite programme will bring a step change in capability for operational meteorology. The programme will guarantee access to space-acquired meteorological data until, at least, the late 2030s. MTG will continue visible and infrared imagery and infrared and ultraviolet/near infrared sounding observations from geostationary orbit that will result in three-dimensional information on humidity, temperature, and wind to support Nowcasting.

The series will comprise six satellites: four MTG-I imaging and two MTG-S sounding satellites. The two types will be positioned over the same longitude in their geostationary orbits. The sounding element will also host the Copernicus Sentinel-4 air quality observatory (see below). The first MTG satellite is planned to be launched around 2020.

The EUMETSAT Polar System (EPS) is Europe's first polar-orbiting operational meteorological satellite system, and it is the European contribution to the Initial Joint Polar-orbiting Operational Satellite System (IJPS), a joint effort between EUMETSAT and NOAA. EUMETSAT's polar-orbiting satellites (i.e. MetOp satellites) carry a set of state-of-the-art sounding and imaging instruments that offer improved atmospheric sounding capabilities to both meteorologists and climatologists.

The EPS Space Segment includes three successive MetOp satellites and is being developed and procured in cooperation between ESA and EUMETSAT. MetOp-A was launched in 2006 and MetOp-B in September 2012. In April 2013, MetOp-B replaced MetOp-A as the EUMETSAT prime operational polar-orbiting satellite, following the end of its commissioning period. The combination of instruments onboard MetOp satellites have remote sensing capabilities to observe the Earth during day- and night-time hours, as well as under cloudy conditions.

The follow-on EUMETSAT Polar System is now under way to replace the current satellite system in the 2020 timeframe and to contribute to the IJPS. Started in 2005 under the name of Post-EPS, the activities have been endorsed in 2012 under a new programme called MetOp Second Generation (MetOp-SG). Comprising six MetOp Second Generation satellites in total, this programme is based on pairs of satellites that carry different packages to deliver complementary meteorological information. The A series of satellites will be equipped with atmospheric sounders as well as optical and infrared imagers, while the B series focuses on microwave sensors. In addition, the A series will carry the Copernicus Sentinel-5 instrument on behalf of the European Commission. Each satellite will be launched separately. It is envisaged that the first A satellite will be launched in 2021, followed by the first B satellite in 2022.

5 The Sentinel Missions

Copernicus is the world's most ambitious Earth observation programme. It is Europe's answer to global needs, allowing to manage the environment, to mitigate the effects of climate change and to ensure civil security Copernicus is led by the European Union, with ESA being the coordinator of the space component. The Sentinels form the dedicated part of the Copernicus Space Component. The Sentinel missions feature constellations of two satellites to provide robust datasets for operational services.

Sentinel-1 This is a polar-orbiting, all-weather, day-and-night mission featuring a C-band Synthetic Aperture Radar (SAR) supporting Copernicus operational services for land, marine, and risk monitoring. The first satellite, Sentinel-1A, was launched in April 2014, and Sentinel-1B in April 2016.

Sentinel-2 This is a polar-orbiting, multispectral high-resolution imaging mission for Copernicus land monitoring to provide imagery of vegetation, soil and water cover, inland waterways, and coastal areas. Sentinel-2 will also provide information for emergency services. The first satellite, Sentinel-2A, was launched in June 2015, and Sentinel-2B in March 2017.

Sentinel-3 This mission is composed of two operational satellites with a one day (land) and a two day (ocean) revisit time. Each satellite will carry an Ocean and Land Colour Instrument (OLCI), a Sea and Land Surface Temperature Radiometer (SLSTR) and a microwave payload including a SAR Radar Altimeter (SRAL) and a two frequency Microwave Radiometer (MWR). This mission will support activities in the field of topography, sea and land surface temperature, ocean carbon, and land colour. The first satellite, Sentinel-3A, was launched in February 2016. Sentinel-3B will be launched in 2018.

Sentinel-4 is a payload devoted to atmospheric monitoring that will be embarked on an MTG-S satellite. Ultraviolet and near infrared spectrometers will be used to continuously monitor air pollution and its precursors from North Africa to northern Europe at a $6 \times 6\ km^2$ spatial and one hour temporal resolutions.

Sentinel-5 Precursor is a polar sun-synchronous mission scheduled to launch in 2015 to reduce data gaps between Envisat and Sentinel-5 for global atmospheric composition measurements including ozone, aerosols and atmospheric pollution precursors. The instrument is a follow-on to the Ozone Monitoring Instrument (OMI) flying on the NASA's Aura satellite with additional channels in the near infrared. Sentinel-5P was launched 13 October 2017.

Sentinel-5 is a payload that will monitor the atmosphere from polar orbit aboard a MetOp-SG A satellite. It will continue the atmospheric composition data collected from Envisat and employ ultraviolet and infrared spectrometers that will the measure profiles and column amounts of trace gases and aerosols important in atmospheric chemistry and climate.

Sentinel-6 carries a radar altimeter to provide high-precision and timely observations of the topography of the global ocean. This information is essential for the continued monitoring of changes in sea level, a key indicator of climate change. It is also essential for operational oceanography. Sentinel-6 builds on heritage from the Jason series of ocean topography satellites and from ESA's CryoSat mission. Sentinel-6 is designed to complement ocean information from Sentinel-3.

6 Outlook

ESA's Earth science program has been steadily growing since the 1990's. Distributing high-quality data and building on continuous technological progress driven by the computer and internet revolution, the "free & open" data policy has been the ultimate accelerator in guaranteeing the maximum exploitation of ESA data.

Copernicus has kicked off a new era of operational Earth observation, leaving the initial "pure science and technology" phase behind and meeting with huge user interest: By November 2017, more than 109.000 users have registered on the Copernicus open access hub, and more than 40 PB of Copernicus data have been downloaded.

Currently Earth Observation is experiencing major trends and paradigm shifts, such as exponential growth in availability of data, the impact of big data and cloud processing, the entrance of new private and institutional actors into the field, the multiplication of national initiatives in Earth Observation, as well as the emergence of venture-capital funded micro-satellite constellations. In addition, advances in on-board technologies are constantly improving spatial resolution and measurement accuracies.

ESA is taking adequate measures to stay on top of these developments. With its suite of missions and programmes, ESA is devoted to bringing the benefits of space based Earth observation to citizens in Europe and worldwide.

Wolfgang Rathgeber is an Electrical Engineer by education. He joined DLR in 1999 as a doctoral student, and he obtained his Ph.D. degree at the University of Karlsruhe with a thesis on Remote Sensing. He went on several secondments from DLR, including the European Space Policy Institute (ESPI) in Vienna and ESA Headquarters in Paris. He then joined ESA as a Programme Coordinator in the Directorate of Earth Observation Programmes in Frascati, Italy.

From Art Space to Outer Space—The Experiment ARTSAT

Richard Kriesche

Artsat is in its last consequence a complete artwork of a new style, in definition the orbit as cultural area. Representative for the citizens of the world an artist shook the cosmonaut's hand in space via electronics. The cosmonaut on his part sent after one full orbit a message to earth. A piano and a robotics were conducted by this message.

Under this aspect of complete artwork artsat goes on with the artistic tradition to bring science, technics, technology and culture together again nearer with the aim to arise art to a norm of workaday.

After 30 years of soviet space flight there was the first time an art project involved into the space program, and so the specific part of Austria in the dialogue with the nations of earth should be shaped out. So artsat got a symbolic role: the inescapable approach of the nations on a common cultural fundament after militarisation, after commercialisation we started with artsat the culturalisation of space.

1 Implementation

6th October 1991 11 CET: The radio contact to the public studio in Graz Image 10 with the space station MIR was compounded, during the MIR was just above the Soviet capital. After a short introduction the artist Richard Kriesche shook symbolicly the hand of the Austrian cosmonaut Image 9. This hand appeared Franz Viehboeck a short time later on a video screen on the space station MIR.

R. Kriesche (✉)
Kulturdata-Graz, Trauttmansdorffgasse 1, 8010 Graz, Austria
e-mail: richard@kriesche.eu

C. Brünner et al. (eds.), *Satellite-Based Earth Observation*,
https://doi.org/10.1007/978-3-319-74805-4_2

From this moment on it took the space station about 60 min until it reached the radio reception area of Graz. During this time a multimedia-meditation with live performances to the theme cultural area space took place, with the Danube waltz as nonstop-background music Images 1, 2, 3 and 4. The time of the orbit was visual implemented with a grave of a burrow as quasi imaginary time sculpture.

As soon as Graz came into the reception area of the space station, the Austrian cosmonaut contacted the public through the amateur radio facility AREMIR. The cosmonaut's message modulated and alienated the Danube waltz like an imaginary conductor's hand. Simultaneous acoustic parameters from the alienate Danube waltz were won and recorded on a computer. The encrypted acoustic message was visualised on a special primed mute piano, Image 5 as if the imaginary hand of the cosmonaut would play the piano.

After the space station left the reception area Graz after a few minutes, the recorded data were accepted by a robot, Image 6 which was built of in front of the public studio. This robot welded the encrypted message in a high-grade steel plate with a diameter of about 3.5 m.

The amateur radio equipment AREMIR from the Austrian payload enabled the realization of this art project.

Image 1 Shaking hands with the cosmonaut

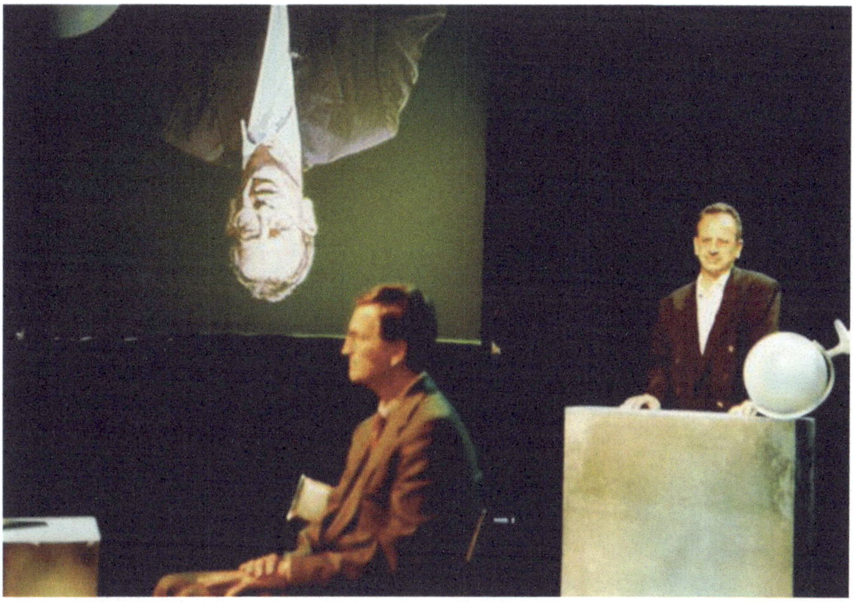

Image 2 The tv-station while transmitting the virtual handshake

Image 3 The cosmonaut awaiting the handshake, 1.1

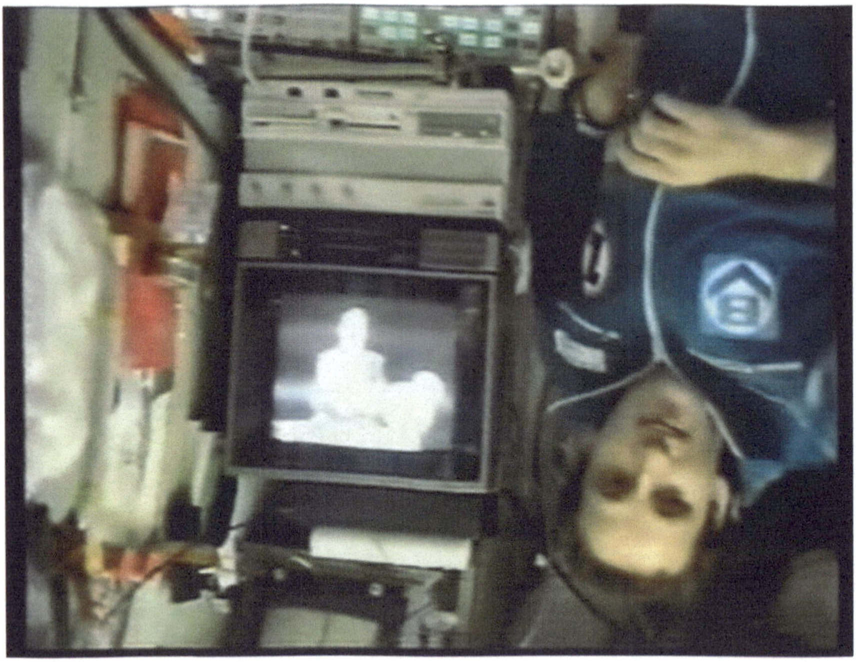

Image 4 The cosmonaut awaiting the handshake, 1.2

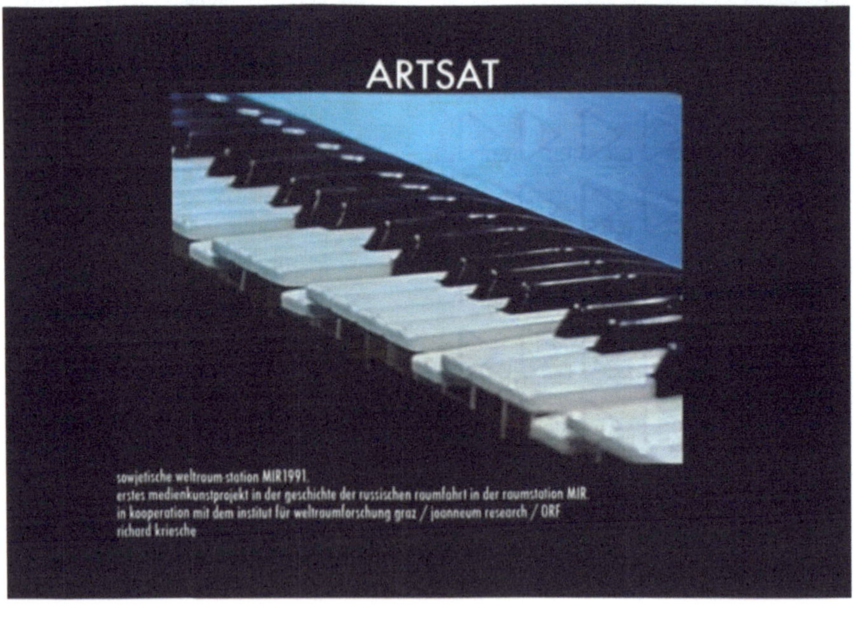

Image 5 The message of the cosmonaut being transmitted into the special prepared piano in the tv-station

Image 6 The welding robot writing the message of the cosmonaut onto a stellplate with 300cm diameter

2 Results

On the 3rd of April 1992 a selected round assembled on the Grazer Schlossberg around the artwork: Image 7. Next to the Federal Minister of Science and Research, the Styrian head of the provincial government and the mayor of Graz several other politician, representatives of the AUSTROMIR-project management and many onlookers came to the exposure of the steel-plate, in which the welding robot had shrink-wrapped the encoded message of the first Austrian scientific cosmonaut.

The glistening steel-rondeau on the Schlossberg, Image 8 which had a special symbolic worth for the "space city Graz", was not the last permanent memory of the project ARTSAT: the same data the welding robot got, were provided to nine Austrian computer musicians. The compositions were saved on CD. This CD is a part of the "ARTSAT-Package", in which all to the project pertaining artistic activities were documented.

The "dusty" surface of the ARTSAT sculpture follows from the visitors' interaction by walking over it.

Image 7 The inauguration of the steelplate at the top of the city's Schlossberg

Image 8 The steelplate

3 Technical Characteristics

The event of "assistance" could be realized with the help of the experiment VIDEOMIR. The video image of the artist's hand in Graz was broadcasted from the public studio of the Austrian broadcast over a radio link of the ORF to the AUSTROMIR-information center on the Institute for Communications Engineering and Wave Propagation at the Technical University Graz. There the video signal was transmitted with the communication satellite EUTELSAT-I-F2 to the radio control

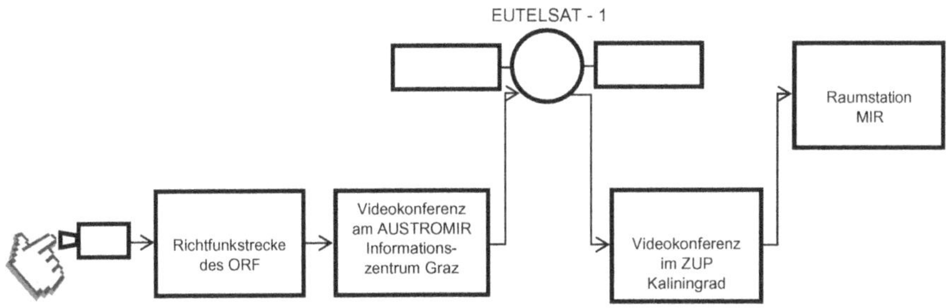

Image 9 Richard Kriesche shook symbolicly the hand of the Austrian cosmonaut

center in Kaliningrad near Moscow. Over this radio system of the radio control center finally the transmission to the space station MIR occured, where the hand appeared on a video screen (Image 9).

When the space station MIR was above Graz, the Austrian cosmonaut contacted the public with the amateur radio facility AREMIR. The cosmonaut's message modulated and alienated the Danube waltz like an imaginary conductor's hand. Simultaneous acoustic parameters from the alienate Danube waltz were won and recorded on a computer. The encrypted acoustic message was visualised on a special primed mute piano, as if the imaginary hand of the cosmonaut would play the piano.

After the space station left the reception area Graz after a few minutes, the recorded data were accepted by a robot, which was built of in front of the public studio. This robot welded the encrypted message in a high-grade steel plate with a diameter of about 3.5 m. This process was projected over video on a big screen in the public studio. Current cross fadings of two camera images were made. One camera was on the arm of the weld robot and the second was mounted on a stand in front of the robot (Image 10).

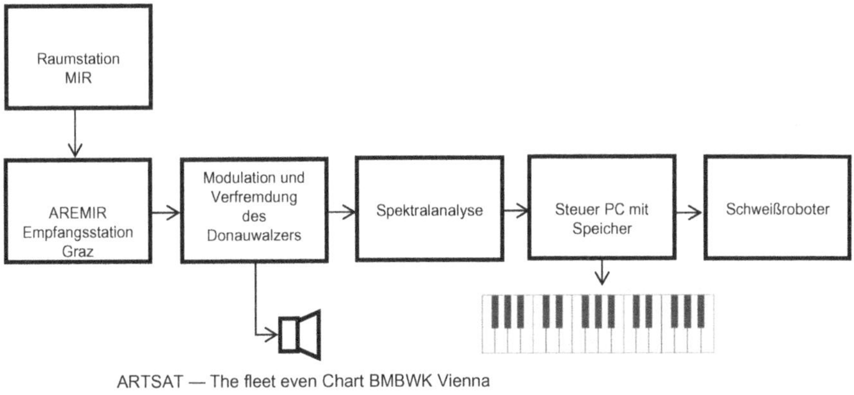

Image 10 The radio contact to the public studio in Graz

4 Experimenters

Prof. Richard Kriesche, Graz
Technics: o. Univ. Prof. Dipl. Ing. DDr. Willibald Riedler (institute manager) Univ. Prof. Dr. Otto Koudelka
Dipl. Ing. Dr. Gerhard Graber (project manager)
all: Institute for Communications Engineering and Wave Propagation, Technical University Graz

Richard Kriesche was born in 1940 in Vienna. 1958–1963 studies at the academy for visual arts Vienna and the university of Vienna. 1963–1990 professor at the htl graz. 1988–1991 assistant professor at the academy for applied arts, Vienna. 1991–1995 fulltime professor at the university for arts and design, Offenbach, germany. 1995/1996 guest professor at the école superieure des beaux arts, paris. 1997 expert of the „council of europe's cultural policy and action division". 1999/2000 independent expert for „policy development in the cultural sphere", resp. for „culture, audiovisual policy and sport" of the European commission. 1995–2005 cultural department of the styrian government. 1984 founder and since 1997 head of "kulturdata".

Part II
Strategic, Technical and Economic Aspects of Satellite-Based Earth Observation

An Introduction

Hannes Mayer

Earth observation (EO) is the gathering of information about the planet Earth's physical, chemical and biological systems by using remote sensing technologies supplemented by earth surveying techniques, encompassing the collection, analysis and presentation of data[1].

Earth observation encompasses different types of sensors (e.g., Synthetic Aperture Radar, Laser Imaging Detection and Ranging, Optical and multispectral) as well as platforms (e.g., satellites, aircraft, and Unmanned Aerial Vehicles) and enables monitoring and modeling geohazards over regions at different scales where ground observation may not be possible due to physical and/or political constraints. Earth observation can provide high spatial, temporal and spectral resolution, stereo-mapping and all-weather-imaging capabilities, but not by a single satellite at a time. Improved satellite and sensor technologies, increased frequency of satellite measurements, and easier access and interpretation of EO data have all contributed to the increased demand for satellite-derived Earth observation data. Earth observation, combined with complementary terrestrial observations and physical models, have been widely used to monitor geohazards, revolutionizing our understanding of how the Earth system works[2].

Although EO techniques were first used for military and security applications in the Cold War, they have become available for a wide range of applications and users over the past few decades[3].

[1]International Journal of Applied Earth Observation and Geoinformation—Aims and Scope. Elsevier. https://www.journals.elsevier.com/international-journal-of-applied-earth-observation-and-geoinformation. Accessed 29 Oct 2017.
[2]Tomás [1].
[3]Ibid.

H. Mayer (✉)
Competence Centre for Space Law and Space Policy, Institute of Canon Law,
Karl-Franzens-University Graz, Graz, Austria
e-mail: hannes.mayer@uni-graz.at

C. Brünner et al. (eds.), *Satellite-Based Earth Observation*,
https://doi.org/10.1007/978-3-319-74805-4_3

Satellite-based Earth observation has many aspects. The launch and operation of satellites is a technical challenge. Satellite observation can be very beneficial for our modern society. The launch and operation of satellites is not only very costly, it can also be highly profitable, concerning launch as well as operation of satellites. The use and exploitation of satellite data is especially important economically. This importance is even greater: Space applications, especially space-based Earth observation, had and continues to have military importance.

Respective data can be derived from proper military satellites as well as procured externally. There are several examples of military and security related space applications. The United States has been and continues to be the largest user of security related space applications. Apart from her large fleet of dedicated military spacecraft —ranging from small (experimental) satellites to large optical reconnaissance satellites of the KH-series to reusable military spaceplanes such as the Boeing X-37B, the United States is also the largest user of space-based assets for military purposes.

This leads us to the whole issue of dual-use applications, meaning space applications which can be used for military and civilian purposes[4]. Earth observation actually is one prime example for civil-military applications.

Apart from that, it has to be noted that space activities in Europe and beyond are often conducted in international and multinational cooperation. Among the examples is the Copernicus program under the auspices of the European Space Agency and the European Union—as mentioned in the keynote address.

Unmanned aerial vehicles are another aspect to be dealt with. Although they are —of course—not to be considered spacecraft, there are certain similarities and analogies to be drawn and certain overlaps. Large UAVs are often controlled via satellite communication. UAVs can be an alternative or a supplement to satellites. They can be used for Earth observation, be it agricultural monitoring, intelligence, surveillance, target acquisition, reconnaissance or even communications.

For matters of terminology: The term remote sensing is often used to refer to satellite-based remote sensing, but sometimes used more broadly for observations using any form of remote sensing technology, including airborne sensors and even ground-based sensors such as cameras. The probably least ambiguous term to use for satellite-based sensors is satellite remote sensing.

References

1. Tomás R, Li Z (2017) Earth observations for geohazards: present and future challenges. http://www.mdpi.com/2072-4292/9/3/194. Accessed 11 Jan 2018
2. Soucek A (2011) Outer space—a fascinating issue. In: Brünner C, Soucek A (eds) Outer space in society, politics and law. Springer, Vienna

[4]Soucek [2].

Hannes Mayer works at the Competence Centre for Space Law and Space Policy as well as the Institute of Canon Law of Karl-Franzens-University Graz. He serves as a member of the Austrian delegation to the United Nations Committee on the Peaceful Uses of Outer Space (UNCOPUOS). He is a member of the European Centre for Space Law, the Space Generation Advisory Council's Space Law and Policy Project Group, the Austrian Space Forum and the History Committee of the International Academy of Astronautics (IAA). He is a board member of the Diocesan Committee of Catholic Organizations Styria and a member of the European Academy of Religion. He has (co-)authored several publications on outer space and is a member of the editorial staff of the Theologicum Magazine.

Strategic Aspects from the European Union's Point of View

Peter Breger

1 Introduction

With the adoption of its Space Strategy for Europe[1] in October 2016 as part of its Space Policy, the European Union has reaffirmed its role in Space, as set out explicitly in Article 189 of the Treaty for the Functioning of the European Union[2] (TFEU). The strategy advocates to use space-related technology to tackle some of the most pressing challenges today, such as fighting climate change, helping stimulate technological innovation, and providing socio-economic benefits to its citizens. Space technology, data and services have become indispensable in the lives of Europeans. We rely on them when using mobile phones and car navigation systems, watching satellite TV, and withdrawing money. Satellites also provide immediate information when disasters, such as earthquakes, forest fires or floods strike, enabling better coordination between emergency and rescue teams.

[1]Communication from the Commission to the European Parliament, the Council, the European Economic and Social Committee and the Committee of the Regions, Space Strategy for Europe, COM/2016/0705 final, 26/10/2016.
[2]Consolidated version of the Treaty on the Functioning of the European Union, OJ C 326, 26.10.2012, pp. 47–390.

P. Breger (✉)
DG for Internal Market, Industry, Entrepreneurship and SMEs,
European Commission, 1049 Brussels, Belgium
e-mail: peter.breger@ec.europa.eu

C. Brünner et al. (eds.), *Satellite-Based Earth Observation*,
https://doi.org/10.1007/978-3-319-74805-4_4

2 Earth Observation in the EU Space Programme

The European flagship programmes and initiatives

- Copernicus[3] for Earth Observation and Monitoring,
- European Global Navigation Satellite Systems of Galileo[4] and EGNOS,
- Space Surveillance and Tracking Initiative,[5]
- Space Research as part of the EU Research Framework Programme Horizon 2020,[6] together provide the European Union's Space Policy with the necessary technical capacity to meet these objectives.

Copernicus, as the Union's Earth Observation (EO) and monitoring programme, enables the Union at political level to fulfill its responsibilities to protect Europe's people and assets, including the environment they live in, through evidence-based policy making. It helps Europe prepare for crises, security risks and natural or man-made disasters, and contributes to strengthening the Union's role as a global soft power. Over the last few years, Copernicus has also become an important tool for economic development and a driver for the digital economy.[7] Being a key ingredient for maximizing the socio-economic value of Earth observation, the legislator has adopted a full, free and open data policy.[8] This latter decision has not been taken lightly, in recognition that a careful balance needs to be found between user needs and the interests of security, a balance which requires continuous examination and evaluation.

[3]Regulation (EU) No 377/2014 of the European Parliament and of the Council of 3 April 2014 establishing the Copernicus Programme and repealing Regulation (EU) No 911/2010, OJ L 122, 24.4.2014, pp. 44–66.

[4]Regulation (EU) No 1285/2013 of the European Parliament and of the Council of 11 December 2013 on the implementation and exploitation of European satellite navigation systems and repealing Council Regulation (EC) No 876/2002 and Regulation (EC) No 683/2008 of the European Parliament and of the Council, OJ L 347, 20.12.2013, pp. 1–24.

[5]Decision No 541/2014/EU of the European Parliament and of the Council of 16 April 2014 establishing a Framework for Space Surveillance and Tracking Support, OJ L 158, 27.5.2014, pp. 227–234.

[6]Regulation (EU) No 1291/2013 of the European Parliament and of the Council of 11 December 2013 establishing Horizon 2020—the Framework Programme for Research and Innovation (2014-2020) and repealing Decision No 1982/2006/EC, OJ L 347, 20.12.2013, pp. 104–173.

[7]See both the Space Strategy for EuropeSpace strategy for Europe and Communication from the Commission to the European Parliament, the Council, the European Economic and Social Committee and the Committee of the Regions, EU Space Industrial Policy Releasing the Potential for Economic Growth in the Space Sector, COM/2013/0108 final, 28/2/2013.

[8]Commission Delegated Regulation (EU) No 1159/2013 of 12 July 2013 supplementing Regulation (EU) No 911/2010 of the European Parliament and of the Council on the European Earth monitoring programme (GMES) by establishing registration and licensing conditions for GMES users and defining criteria for restricting access to GMES dedicated data and GMES service information, OJ L 309, 19.11.2013, pp. 1–6.

The objective of Copernicus is to provide timely and reliable geo-information, as well as forecasts of the atmosphere composition and ocean physical state. To this end, it combines (see Fig. 1) observations from satellites (both dedicated Copernicus satellites as well as other contributing satellite missions) and in situ measurements (from ground, air, ship or buoy based sensors). Combining all these information sources, Copernicus services transform raw observation data into value-added geo-information products, which can then be used directly by decision-makers. These products are structured around 6 thematic areas: Land Monitoring; Marine Environment Monitoring; Atmosphere Monitoring, Emergency Management, Security and Climate Change.

Through its public services, Copernicus enables public authorities at European, national and regional, right down to local level, to fulfil their task of both policy making and administration. At the same time, Copernicus is also a driver for economic growth. As a public good, the products must act as a foundation of information which can be used in turn by businesses to develop tailor-made product for a wider clientele.

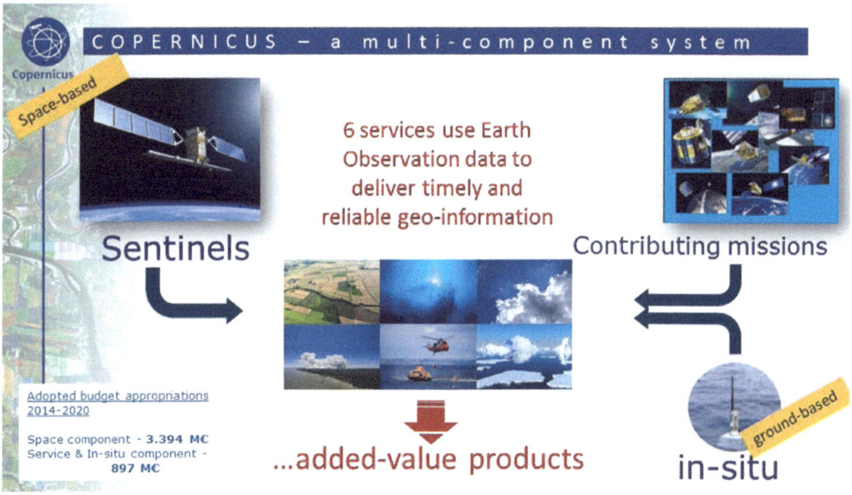

Fig. 1 Copernicus architecture as a multi-component system. Sentinel pictures: © ESA available here: https://www.esa.int/spaceinimages/Images. In-situ picture: © Ifremer. Land—iStock: © Dariusz Paciorek. Marine—iStockphoto: © MyOcean. Atmosphere: © FDC. Climate change—Fotolia: © z576. Emergency—Fotolia: © Tad Denson. Security—Fotolia: © scazza

3 The Copernicus Data Acquisition—from Space and in Situ Sensors

A dedicated set of Sentinel satellites provides Copernicus with space data carefully tailored to meet European user needs. Developed and constructed in partnership with the European Space Agency as technical architect, the Sentinel missions provide capacities as follows:

- The two satellites Sentinel 1 (A&B) provide all-weather, day and night radar imaging at 9–40 m resolution with a 6-day revisit time at the equator;
- The two satellites Sentinel 2 (A&B) provide polar-orbiting, multi-spectral optical high-resolution imaging at 10-60 m resolution with a 5-day revisit time;
- Sentinel 3 (A) is an optical imaging and altimeter mission, monitoring sea and land parameters with a global focus at 300–1200 m resolution with a revisit time of less than two days;
- The future Sentinel 4 payload on the MTG-S satellite adds an atmospheric chemistry monitoring capacity, while
- Sentinel 5p (launched in November 2017) is a mission designed to reduce the data gaps of atmospheric chemistry data between the previous Envisat mission and future Sentinel 5 satellites.
- Sentinel 5 will provide atmospheric chemistry data at 7.5–50 km resolution globally, with a daily revisit time.
- Sentinel 6 will provide a continuation of the current Jason missions, as an ocean altimeter reference mission, measuring sea-surface heights globally.

A deployment schedule is depicted in Fig. 2, and shows the successful launches to date with five satellites in orbit by end of 2017.

A key strategic aspect of the deployment schedule is its continuity up to 2030—all satellites depicted are financed in their construction from the current Copernicus programme budget appropriations, thus providing operational predictability. This allows users, particularly in the downstream commercial sector, to align their commercial operations, thus enabling long-term business plans. Further to this data source from space, Copernicus benefits from a wealth of space data from so-called Contribution Missions, be they national or of commercial nature. With a growing interest by the international community in cooperating on exchange of data with the EU in Copernicus, several administrative agreements are already in place incorporating data from satellites owned by third countries.

Copernicus also relies heavily on in situ data, meaning observation data from ground-, sea-, or air-borne sensors, reference and ancillary data licensed for use in Copernicus. These enable Copernicus products to be validated and calibrated, and to provide reliable information services rather than only space data. This data is integrated in the programme in two tiers, firstly by incorporating tailored in situ data for each Copernicus service level, but also by providing a cross-cutting access coordination across all services by the European Environmental Agency (EEA).

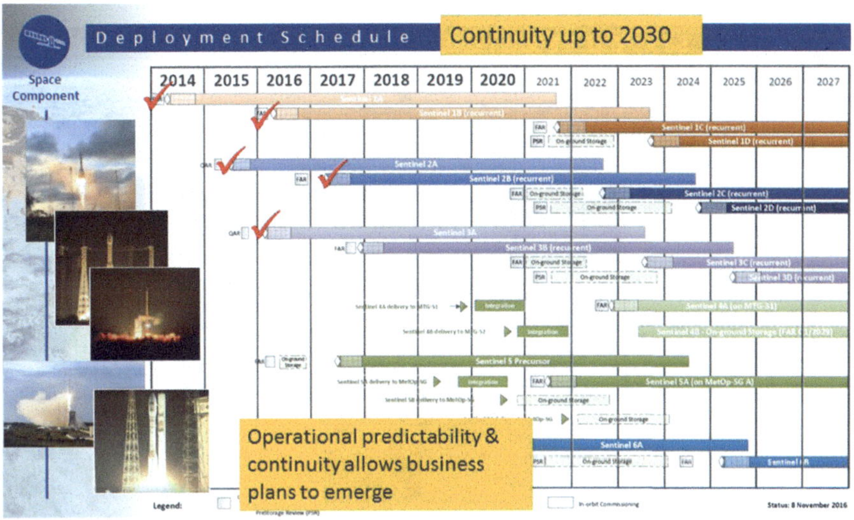

Fig. 2 Deployment schedule for sentinel satellites—with satellites shown (tick mark) to have been launched and in orbit. Launch pictures: © ESA available here: https://www.esa.int/spaceinimages/Images

4 The Copernicus Data Processing and Service Component

With its six Services (see Fig. 3) at the heart of the programme, the range of space infrastructure to be constructed and operated, and non-space data providers to be integrated, Copernicus can be realised only as a partnership of diverse and distributed actors. As a Union programme, Copernicus is coordinated and managed by the European Commission. It is implemented in partnership with the Member States, the European Space Agency (ESA), the European Organisation for the Exploitation of Meteorological Satellites (EUMETSAT), the European Centre for Medium-Range Weather Forecasts (ECMWF), EU Agencies and Mercator Océan. As regards the Space Component, the European Commission has entrusted ESA to lead the technical coordination, procurement and operations for the Sentinel fleet, and EUMETSAT for the operations and maintenance of the atmosphere and ocean monitoring satellites. As regards the Service component, activities are delegated to competent service operators that manage these, creating operative products and performing development activities as needed to maintain and update the products:

- **Land Monitoring Service**—delegated to the EEA and the DG Joint Research Centre. This Land Service is divided into three main components. The Global Land Service providing a series of bio-geophysical products on the status and evolution of land surface at global scale at mid and low spatial resolution. Production and delivery of the parameters take place in a timely manner and are

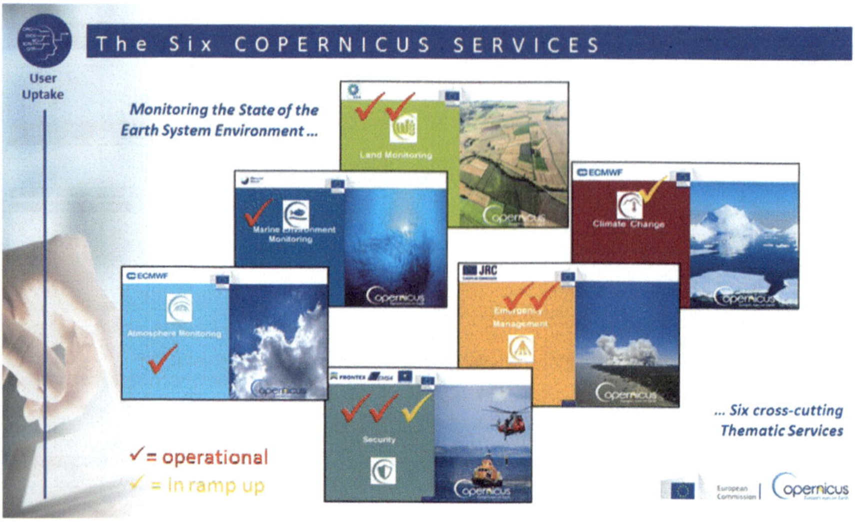

Fig. 3 The six Copernicus services—addressing the three earth system domains and cross-cutting thematic domains. Land—iStock: © Dariusz Paciorek. Marine—iStockphoto: © MyOcean. Atmosphere: © FDC. Climate change—Fotolia: © z576. Emergency—Fotolia: © Tad Denson. Security—Fotolia: © scazza

complemented by long-term time series of data. The products are used to monitor vegetation, the water cycle and the energy budget. The pan-European component provides information about land cover and land use (LC/LU), changes to these and land cover characteristics. The latter includes information about imperviousness, wetlands, permanent water bodies and vegetation. The local component focuses on different hotspots, i.e. areas that are prone to specific environmental challenges and problems. This includes detailed LC/LU information for the larger EU cities (Urban Atlas), riparian zones along European river networks and NATURA 2000 sites. It will also include maps of coastal areas for their management.

- **Emergency Management Service**—delegated to the DG Joint Research Centre, with operational 24/7 coordination being performed by the DG ECHO Emergency Response Coordination Centre (ERCC) as the single entry-point for activating the mapping service. The service provides information for emergency response in relation to different types of disasters, such as floods, fires or earthquakes. The mapping service delivers reference, delineation and grading maps. Two further components are the European Flood Awareness System (EFAS) and the European Forest Fire Information System (EFFIS), both devoted to supporting Member States authorities with risk assessments and alerts.

- **Marine Environment Monitoring Service**—delegated to Mercator Océan. It provides a global model of the state of ocean, describing currents, sea-surface temperature and sea-levels at global and regional levels. A seamless model covers the full global ocean system, and regional seas under the strong influence of tidal dynamics and coastal boundaries. These data are vital for understanding the ocean and many activities such as planning of shipping routes and fisheries.
- **Atmosphere Monitoring Service** and **Climate Change Service**—both delegated to the European Centre for Medium-Range Weather Forecasts (ECMWF). The Atmosphere service provides information on the atmospheric composition, such as real-time analyses and forecasts of green-house gases and air pollution on daily or even more frequent basis. These information products are accompanied by consistent reanalysis products. The Climate Change Service has developed a climate change data infrastructure, acting as an authoritative source of climate data, offering reanalysis products, as well as toolboxes for projections of climate change.
- **Security Service**—delegated to the European Border and Coast Guard Agency (FRONTEX), the European Maritime Safety Agency (EMSA) and the European Union Satellite Centre (EUSC). FRONTEX main objectives in the area of **border surveillance** are to reduce the number of illegal immigrants entering the EU undetected, to reduce the death toll of illegal immigrants by rescuing more lives at sea and to increase internal security of the European Union as a whole by contributing to the prevention of cross-border crime. In the area of **maritime surveillance**, the overall objective of the European Union is to support Europe's maritime security objectives and related activities in the maritime domain. The corresponding challenges mainly relate to navigation safety, support to fisheries control, combating marine pollution, and law enforcement. A third component of the Security Service assists the Union in its role as a global actor **supporting actions external to the European Union**. In this context, the major objective of the EU is to assist third countries in a situation of crisis or emerging crisis and to prevent global and trans-regional threats having a destabilising effect.

After nearly three years of operation, the Copernicus infrastructure and services are set up as planned and are well established. In this short space of time, Copernicus has not only become the world's largest Earth Observation programme, but, by incorporating Copernicus Services knowledge generation into its architecture, it has become a global centre of Earth Observation related scientific expertise.

5 Reaching the User—Distribution and Uptake

Establishing an internal market for innovative and space based application has been and remains a primary goal of a European Space Policy (Fig. 4). In this respect, specific objectives have to be met to address all the elements of a space-based value chain, covering Data Generation, Data Distribution, the subsequent Processing steps, and uptake with integration into workflows at the end-user.

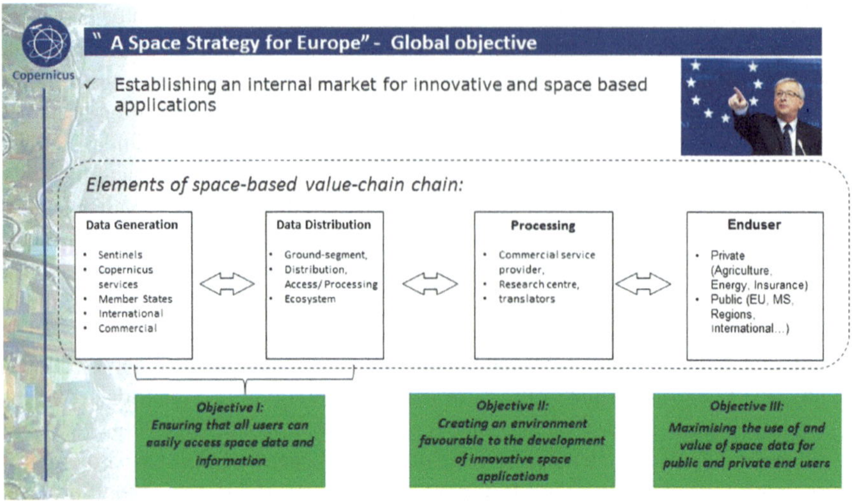

Fig. 4 Elements of a space based value chain in Copernicus. Jean-Claude Juncker—Fotolia: © European Union

The first objective therefore must be to ensure equality of access to space data and information. The insistence on a full, free and open data policy and public access points to discover, view and download Copernicus data and products thus plays a central role in opening up this information to all users, both large and small.

Creating an environment favourable to the development The first objective therefore must be to ensure equality of access to space data and information. The insistence on a full, free and open data policy and public access points to discover, view and download Copernicus data and products thus plays a central role in opening up this information to all users, both large and small.

of innovative space applications has been identified as another prime objective, and has received additional attention over the last year. This, in turn, has been recognised as an important pre-requisite to maximising the use of and value of space data for public and private end users.

The European Space Strategy is clear about promoting the uptake of Space services and data, and explicitly states that "*the Commission will:*

- *facilitate the use of Copernicus data and information by strengthening data dissemination and setting up platform services, promoting interfaces with non-space data and services;*
- *stimulate the development of space applications with a greater involvement of new actors from different domains;*
- *together with Member States and industry, promote the efficient and demand-driven use of satellite communications, so as to foster ubiquitous connectivity in all Member States."* [Space Strategy for Europe, Section 1.1, page 4]

The recent developments undertaken by the European Commission, in restructuring its Copernicus funding and support instruments, have been designed to meet these challenges head on. Weaknesses had been identified in the prevailing distribution and access systems, in failing to be sufficiently open and user friendly, inadequately incorporating both space data and information from the services in an equal manner, as well as failing to incorporate third party data in an attractive processing environment. This last desired feature should appeal to users to bring their own processing to the data, and thereby bring about the paradigm change from "bringing data to the user" to "bringing users to the data".

The rapidly rising amounts of data available from Copernicus clearly illustrate the need for such a paradigm change—with a current data archive of some 8.5 Petabytes, this is projected to reach some 20 Petabytes by 2020 and 43 Petabytes by 2024. Data transfers between Copernicus pick-up points and users are currently too slow to realise required download needs, thus impinging on users' ability to make effective use of Copernicus data and information on large scales. Consequently, the European Commission is focussing its efforts on strengthening traditional distribution routes and launching innovative Data and Information Access Services (DIAS) to bring users close to the data. These steps engage with state-of-the-art solutions to build a digital economy, thereby also addressing a policy priority of the European Union.

Of course, users have to be aware of the availability of data, which in turn will stimulate demand and then innovative ideas. This, however, is realisable only if tackled in cooperation with Member States. It is addressed specifically through the Copernicus Relays (hubs and clusters) and a Framework Partnership Agreement for which calls of interest[9] and calls for proposals[10] have been published and agreements are being forged in 2017. A further focus of activities is directed to stimulate the wider use of Copernicus data and information in agriculture, energy provision, insurance, raw materials and mining sectors, with workshops throughout the year being just one form of engagement with this challenge. Responding to user needs in these different domains is key in all these activities, a rationale which is repeatedly stressed throughout the Space Strategy for Europe.

6 Preparing the Future—Strategic Aspects

Having now reached mid-term of the current Copernicus Programme 2014–2020, it is time to prepare the transition from the current regulation to the next. Turning to the future evolution of Copernicus as the Union's Earth Observation System, the Space Strategy for Europe sets some clear priorities:

[9]http://www.copernicus.eu/sites/default/files/documents/News/Copernicus_Relays_Call_for_EOI.pdf.

[10]http://www.copernicus.eu/news/call-proposals-caroline-herschel-framework-partnership-agreement.

> *"The Commission will address emerging needs related, in particular, to climate change, sustainable development and security and defence."* [Space Strategy for Europe, Section 1.2, page 5]

With this in mind, additional services will be considered for preparing a Commission proposal. An important principle in setting the scope for future action is that *"the private sector will only use and invest in space solutions if users and businesses are confident that services and data provided by Copernicus ... will continue in the future. The Commission therefore confirms its commitment to the stability of the EU space programmes"* [Space Strategy for Europe, Section 1.2, page 5]. As a top priority, the stability of a financing horizon is needed, supporting (enhanced) continuity of current data and services, as well as continuity of the full, open and free data policy for the environmental domain. The following observation needs from Copernicus will be further investigated for their potential as environmental observations:

- Priority 1:
  - Greenhouse gas monitoring, specifically on anthropogenic CO2 emissions for which currently no satellite observations are available
- Priority 2:
  - Monitoring the Polar regions, specifically the Polar/Arctic observations
  - Monitoring agriculture, specifically on parameters which potentially could be addressed through thermal infrared observations
- Priority 3:
  - Mining, biodiversity, soil moisture and other parameters, requiring observations in additional bands, currently not available

Taking a much stronger role on the world stage has been prioritised by the Space Strategy for Europe. In particular "*the Commission will*

- *pursue space dialogues with strategic international partners;*
- *use economic diplomacy and trade policy instruments to assist European companies active in global markets and to address societal challenges;*
- *foster the EU's contribution to international initiatives such as the Group on Earth Observation (GEOSS) and CEOS;*
- *engage with international partners together with the other EU institutions and Member States.* " [Space Strategy for Europe, Section 4, page 12]

It is therefore only timely and most fitting that in October 2017 the Commission took over the Chairmanship of CEOS for 2018, a position which rotates annually between three regions of the world (America, Europe, Asian region), and is held in 2017 by the U.S. Geological Survey.

As emphasised already above, central to the success of Copernicus is the effort to improve the framework conditions for commercial Space activities, i.e. an environment which incentivises the private sector to be more risk - prone and encourages businesses to develop innovative products and services. In this respect, "*the Commission will*

- *step up its efforts to support space R&D activities, in* ***cooperation with Member States and ESA****, and review its strategic approach to boosting the competitiveness of the European space sector."* [Space Strategy for Europe, Section 2.1, page 7]

A further strategic aspect, not solely applicable to Earth Observation, but equally a prerequisite for success, is consolidating Europe's autonomous access to space. Hence *"the Commission will:*

- ***aggregate demand for launch services*** *to provide visibility to industry and reduce implementation costs."* [Space Strategy for Europe, Section 3.1, page 9]

An action identified by the Space Strategy for Europe to ensure this freedom of action and autonomy is that

- ***"Access to the radio frequency spectrum*** *must be guaranteed and protected from interference."* [Space Strategy for Europe, Section 3.2, page 9]

7 Conclusions

Just three years after the adoption of the Copernicus Regulation and the launch of the first Sentinel satellite, Copernicus is producing tangible results which clearly demonstrate the added value of this EU Programme. Copernicus is a great opportunity for Europe, offering a huge potential for innovation, growth and jobs. With Copernicus, the European industry has a unique opportunity to become a leader in a fast-growing market. The Space Strategy for Europe acknowledges this achievement and proposes a range of actions to allow Europeans to fully seize the benefits offered by space, which will guide the preparation of a Commission proposal for future funding.

8 Acknowledgements

This article is a written version of the presentation given at the Symposium "Trends and Challenges of Satellite-based Earth Observation for Economy and Society", and the views expressed are purely those of the writer and may not in any circumstances be regarded as stating an official position of the European Commission. The author would like to thank colleagues and the entrusted entities of Copernicus for their contributions.

Peter Breger is the Deputy Head of Unit for "Copernicus" in the European Commission since July 2013. This unit implements and manages the space infrastructure and six operational services of Copernicus following the adoption of the Copernicus Regulation on 3 April 2014. Dr. Breger has been active in the Commission's Space domain since 2002, both in the precursor of Copernicus in its initial phases, and in implementing the Space theme of the EU Research Framework programme.

Space and Security: Earth Observation Between the Priorities of Civilian and Military Use

Norbert Frischauf

1 Introduction

September 11, 2001 marks the date when the power of Earth imagery became obvious to the general public for the first time. Due to the terror attacks on the World Trade Centers and other sites, all airplanes were banned from the US skies for several days. Consequently it was up to Earth Observation satellites like IKONOS and others to provide exclusive images of "Ground Zero", such as the one depicted in Fig. 1 Launched two years earlier and operated by DigitalGlobe, IKONOS was the first of its kind; a commercial Earth Observation satellite providing images in four visual bands with a resolution as good as 1 m, demonstrated on the right side of Fig. 1 where a zoom-in into the photo clearly shows cars that are driving along the highway on the waterfront.

Figure 2 outlines some key specifications for the first generation of commercial Earth Observation satellites, such as IKONOS and QuickBird. Contrary to the original schedules, IKONOS and QuickBird were not operated for 5–7 years. Instead both provided high resolution pictures at a rate of 900 images per day for a period of 15 and 13 years respectively. If a commercial system is operated 2–3 times longer as originally planned for, the reason for doing so is evident: commercial success. The typical cost for a QuickBird panchromatic image with a resolution of 0.61 m and a

N. Frischauf (✉)
CERN, Geneva, Switzerland
e-mail: norbert.frischauf@cern.ch; norbert.frischauf@oewf.org

N. Frischauf
Austrian Space Forum (OEWF), Vienna, Austria

C. Brünner et al. (eds.), *Satellite-Based Earth Observation*,
https://doi.org/10.1007/978-3-319-74805-4_5

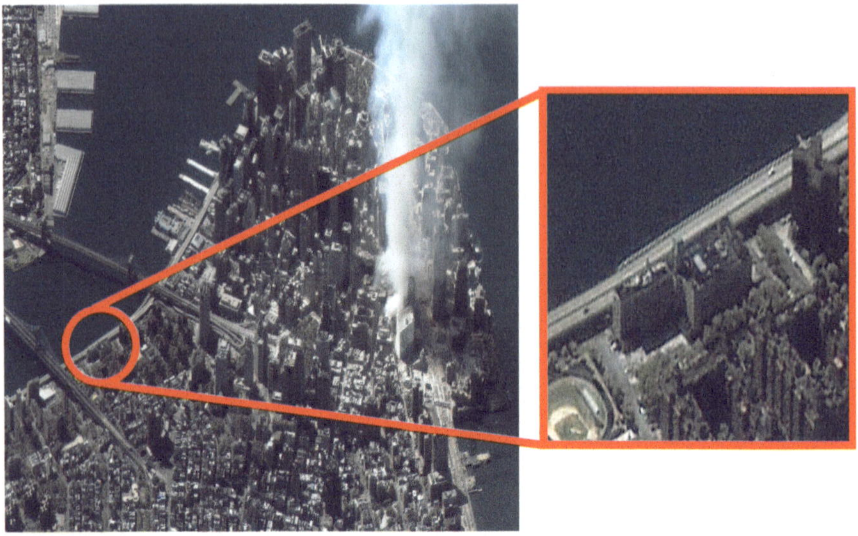

Fig. 1 Downtown Manhattan as seen by the IKONOS satellite on Sept. 12, 2001. *Source* DigitalGlobe (Image taken from: http://mashable.com/2015/05/28/ikonos-satellite-digitalglobe/#ZRADSATRQiqy)

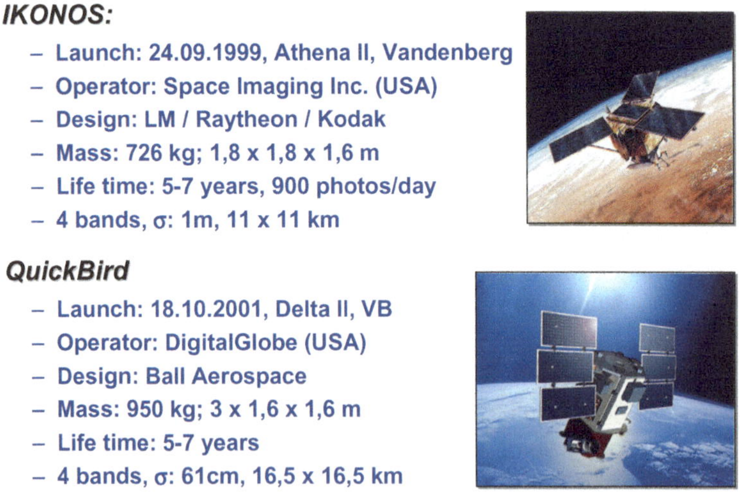

Fig. 2 Specifications of commercial „Spy Satellites", which provide for hi-res photos as seen on Google Maps (Images taken from: http://space.skyrocket.de/doc_sdat/ikonos.htm and https://www.satimagingcorp.com/satellite-sensors/quickbird/)

swath width of 16.5 km delivered within 2–3 days amounted to approx. 24 USD/km^2.[1] When QuickBird finally re-entered the Earth's atmosphere on January 27, 2015 it had made more than 70,000 trips around the planet and captured some 636 million km^2 of high-resolution Earth imagery that contributed to the operator's income statement and to humanity's understanding of our changing planet.[2] Based on this success, both IKONOS and QuickBird have become the forerunner to today's high quality satellite imagery services, which make it possible for every one of us to download pictures with a resolution of up to 30 cm or HD videos of the Earth within a few seconds onto our computer desktops or smartphones.

2 From Science and Earth Observation to "Battleground Space"

Earth Observation was one of the early applications of spaceflight (Fig. 3). When the first rockets were launched into space after World War II, science and reconnaissance were the main drivers. Over time, space saw an ever increasing military utilisation showcased by purposes such as:

- Navigation (GPS)
- Space reconnaissance (especially observation of foreign ICBMs)
- Control of own ICBMs and long range guided missiles (performed by US/USSR)
- Communication
- (SDI)

SDI, the Strategic Defense Initiative was a proposed missile defense system intended to protect the United States from an attack by ballistic strategic nuclear weapons (Intercontinental Ballistic Missiles [ICBMs] and Submarine-Launched Ballistic Missiles [SLBMs]). Announced by US-President Ronald Reagan on 23 March 1983, efforts to develop SDI systems commenced through the 1980s up to 1993 when the political support for SDI collapsed due to the end of the Cold War.[3]

While SDI never became operational, the militarisation of space continued. Already in 1985, the USA had set up the United States Space Command (USSPACECOM) to oversee US space activities. Space assets gave the coalition in Operation Desert Storm (1990–1991) a decisive edge and later operations in the Balkans, Southwest Asia, Afghanistan and Iraq relied heavily on space-based command and control, communications, surveillance and intelligence, navigation and weather systems. In 1992 a new single command responsible for all strategic nuclear forces was formed, the United States Strategic Command (USSTRATCOM). The new command's principal mission was to deter military

---

[1]Huggel et al. [1].

[2]Satellite Imaging Corporation [2].

[3]Baucom [3].

Fig. 3 Image of the pentagon, made by a Corona spy satellite on 25 September 1967. *Source* NRO (Image taken from: https://commons.wikimedia.org/wiki/File:Corona_pentagon.jpg)

attacks, especially nuclear attacks, on the United States and its allies and, if deterrence failed, to employ nuclear forces. Both commands were finally merged in 2002 leading to the creation of the current USSTRATCOM, which is still active today.[4] Its current mission has enlarged to employ tailored nuclear, cyber, space, global strike, joint electronic warfare, missile defense, and intelligence capabilities to deter aggression, decisively respond if deterrence fails, assure allies, shape adversary behaviour, defeat terror and define the force of the future.[5]

Figure 4 depicts the seal of USSTRATCOM, with the history and legacy of strategic and space operations being represented. One of the many symbols in it is the globe, as being viewed from space symbolising the Earth as being the origin and control point for all space vehicles and represents the command's span of operations. The various emblem colours represent the joint character and rigor of the command, whereby the blue command designation band represents the command's air-based and responsive ballistic missile force, agile bomber assets, aerial refuelling, reconnaissance aircraft and airborne command platforms.[6]

[4]U.S Strategic Command [4].
[5]U.S Strategic Command [5].
[6]Wikimedia Commons [6].

Fig. 4 Seal of the United States strategic command. *Source* USSTRATCOM

3 Air and Space Reconnaissance in the Cold War

While reconnaissance had always played an important role in conflict times, the end of World War II and the start of the Cold War gave it a clear prime role, fuelled by the two buzz words "bomber gap" and "missile gap". Both Cold War terms were used in the USA in the 1950s and 1960s for the perceived superiority of the number and power of the USSR's bombers and missiles in comparison with the systems available to the USA. Sparked by ICBM tests performed by the Soviets in August 1957 and the successful launch of Sputnik I in October 1957, the United States began to believe that the Soviet Union possessed superior missile capabilities that directly threatened the continental USA. Moreover, US military and intelligence agencies projected that the Soviet Union would likely improve its missile technology significantly, as well as increase its numbers of nuclear missiles, relative to that of the United States. The Gaither Report, issued in November 1957, gave a comparative analysis of the state of US and Soviet nuclear forces and presented policy proposals. The report argued that US nuclear strategy could no longer be built around its superior strategic bomber force and its destructive capacity, because those could be neutralised by a preventive surprise missile attack.[7]

Quantifying the bomber and missile gap was the task of the hour and consequently efforts were undertaken to obtain reconnaissance data of the presumed large numbers of bombers and missiles hiding behind the iron curtain. The tool of choice to obtain such information was the U-2, a US high altitude reconnaissance aircraft, built by Lockheed Skunk Works, which saw its introduction into service in 1957. With a range of more than 6000 nautical miles, a service ceiling of nearly 85,000 ft

[7]Encyclopaedia Britannica [7].

Fig. 5 USAF Lockheed U-2 Dragon Lady. *Source* USAF (Image taken from: https://en.wikipedia.org/wiki/Lockheed_U-2#/media/File:Usaf.u2.750pix.jpg)

(25,900 m) or more and an endurance of 12 h, the U-2 was believed to be out of range for Soviet radar, interceptors and incoming missiles (Fig. 5).[8]

The shooting down of a U-2 on 1 May 1960 proved that this assumption was wrong, when Francis Gary Powers performed a deep-penetration Soviet overflight crossing the USSR on a planned route starting in Peshawar, Pakistan and ending in Bodø, Norway (Fig. 6).

The CIA designation for this 24th deep-penetration Soviet overflight was Operation GRAND SLAM and what looked like a successful mission for the first four and a half hours into the flight turned out to be a disaster when the Soviets fired three SA-2 missiles at the spy plane over Sverdlovsk. One of these detonated behind the aircraft at 70,500 feet and made the U-2 crash in the USSR. The pilot survived the crash and got captured, the U-2 was not entirely destroyed allowing the Soviets to identify much of its equipment. A public trial starting in August 1960 was set up to disgrace the US and saw Powers being sentenced to three years of prison. He got finally exchanged for Rudolf Abel on 10 February 1962[9]—a first-ever spy exchange, which has been nicely covered in the recent movie "Bridge of Spies" featuring Tom Hanks as the lawyer James B. Donovan, who is entrusted to negotiate the release of Powers in exchange for Abel, a convicted Soviet KGB

[8]Military Factory [8].

[9]CIA [10].

Fig. 6 U-2 "GRAND SLAM" flight plan on 1 May 1960. *Source* CIA [9]

spy held under the custody of the United States, whom he had represented at trial. The name of the film refers to the Glienicke Bridge, which connects Potsdam with Berlin, where the prisoner exchange took place (Fig. 7).

Fig. 7 Poster for the film "Bridge of Spies", released in 2015 (Image taken from: https://en.wikipedia.org/wiki/Bridge_of_Spies_(film)#/media/File:Bridge_of_Spies_poster.jpg)

While the shooting down of Powers' U-2 did not lead to the halt of the utilisation of the U-2 as a reconnaissance aircraft,[10] it did nonetheless push forward the satellite reconnaissance programmes and notably Project Corona, a US strategic reconnaissance satellite programme, which had already started in June 1959.

[10]The current model, the U-2S, received its most recent technical upgrade in 2012. Recently, U-2s have taken part in post–Cold War conflicts in Afghanistan and Iraq and supported several multinational NATO operations. A potential retirement of the U-2 is on hold until 2019 (according to Breaking Defense [11]).

4 Corona: The First US Spy Satellite that Triggered a Spy Thriller Novel and Movie

The Corona program was a series of US strategic reconnaissance satellites produced and operated by the Central Intelligence Agency (CIA) Directorate of Science and Technology with substantial assistance from the US Air Force (USAF). The Corona satellites were used for photographic surveillance of the USSR, the People's Republic of China and other areas beginning on 18 August 1959 and ending on 25 May 1972.[11]

Typical for secret programmes, several other names have been put in use for Corona, such as Discoverer and Samos. Bearing in mind that Corona is only but the first US spy satellite system in a long series of follow-up programmes, several names have emerged for the different US spy satellite systems, such as Argon, Lanyard, Gambit, etc. This plethora of names would make it more difficult to track and sort the different systems, luckily however, there is a common element to all these programmes: the main imaging payload, which is designated as "Keyhole" or "KH". Looking for KH therefore will relatively easily provide for a listing of the US spy satellite systems. In case of Corona, KH-1 up to KH-4, as well as KH-4A and KH-4B were the camera systems in use.

When Corona was launched, CCD systems were still futuristic Science Fiction, the state-of-the-art camera system relied on chemical film—a film that had to be stored, moved forward on a reel, exposed image by image and finally called for processing in a chemical lab. Consequently a Corona mission was complex, as depicted in the picture sequence in Fig. 8.

A typical Corona mission lasted for 19 days. Launched by a THORAD-AGENA booster, the satellite was delivered in a 186 × 280 km polar orbit to perform its reconnaissance objectives. While early satellite featured a mass of 780 kg, later generations like the KH-4B depicted in the right center of Fig. 8 were as heavy as 2000 kg. Since the system relied on chemical film a complicated film transport mechanism had to be used to store and process the photos. In the end the system was required to handle up to 4900 m of film for each of the two cameras, summing up to a total amount of 9800 m of film carried on board.[12] Once the film storage was exhausted, the final part of the mission sequence was initiated, cumulating in a very particular and spectacular feature of every Corona mission; the recovery of the descending Corona capsule by an aircraft, depicted in the lower right corner of Fig. 8. The aircraft to perform this specific task was a US Air Force C-119; typically the capturing of the capsule happened in an altitude slightly lower than 5000 m.

Albeit very complex, 102 out of the 145 missions very successful. In total 860,000 photos were shot, 39,000 film canisters utilised and the covered land area

[11]Ruffner [12].

[12]Brown [13].

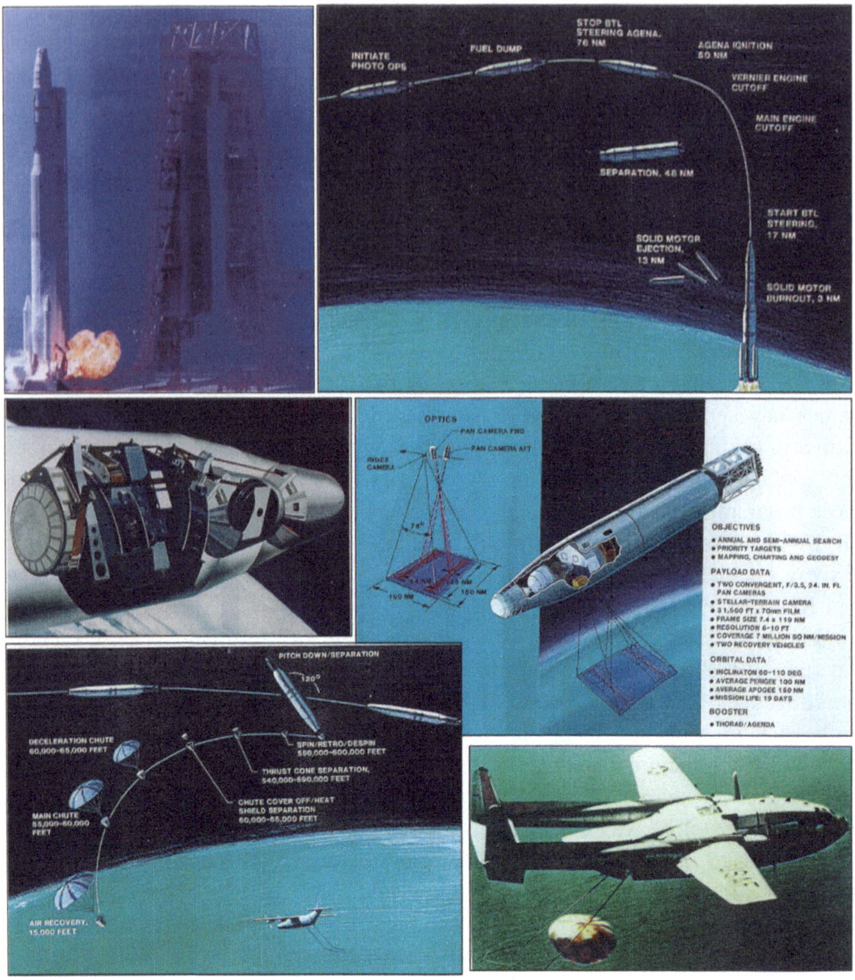

Fig. 8 A typical sequence of a Corona reconnaissance mission. *Source* NRO (Images taken from: http://www.nro.gov/history/csnr/corona/imagery.html, as well as https://commons.wikimedia.org/wiki/File:Kh-4b_corona.jpg, https://en.wikipedia.org/wiki/Corona_(satellite)#/media/File:CORONA_film_recovery_maneuvar.jpg, etc.)

surmounted to nearly 2000 million square kilometers. The resolution of the black and white image was astounding: 1.5–1.8 m were feasible, a typical photo of the Corona programme is depicted in Fig. 3.

The most prominent failure of the Corona programme happened rather early in the programme's lifecycle, when a Corona satellite capsule inadvertently landed near Spitzbergen on 13 April 1959. The event made it into the news, a race between the Soviets and the American developed, who would retrieve as first. While it is possible that Soviet agents may have recovered the capsule (Discoverer II mission,

Fig. 9 Poster for the film "Ice Station Zebra", released in 1968 (Image taken from: https://en.wikipedia.org/wiki/Ice_Station_Zebra#/media/File:Ice_Station_Zebra_(film)_poster.jpg)

which did not fly yet with a camera), it is more likely that the capsule landed in the water and sank[13] Whatever had happened, the 1963 thriller novel "Ice Station Zebra" and its 1968 film adaption were in part inspired by this event. It is said that the US billionaire Howard Hughes was the biggest fan of that movie as historic witnesses like Paul Anka, affirmed that he had watched a personal copy of "Ice Station Zebra" 150 times on a continuous loop in his private hotel suite during the years prior to his death (Fig. 9).[14]

[13]Day (2008) Spysat down! The Space Review.

[14]Chun [14].

5 Argon, Lanyard, Gambit, etc.—Corona's off-springs

If anyone would have been sceptic that spy satellites could deliver, the success of the Corona programme had proven them otherwise. After Corona, spy satellites became the accepted tool of choice for the reconnaissance demands of the Air Force, Army, CIA, NRO and others. Consequently spy satellite activities increased, cumulating into several generations of reconnaissance satellites, one more performant than the other. Starting in 1962 with the 1274 kg massive KH-5 Argon, a programme lasting from 1962 to 1964, encompassing 12 missions of which 6 were successful and cumulating with the nearly 20 ton massive KH-12 Improved Crystal, which saw its first mission in 1990 ever bigger and more massive systems had been launched into orbit (refer to Figs. 10 and 11.)

When looking at Fig. 11, KH-11 Kennan features a striking resemblance with the Hubble Space Telescope (HST), depicted on the left side within Fig. 12. But while the HST is a unique piece of space hardware, its sister system KH-11 features a complete constellation of satellites to provide for a continuous observation of the Earth (right image within Fig. 12).

The latest generation of optical US spy satellites like KH-11 and KH-12 feature resolutions better than 10 cm. These optical systems are nowadays complemented by radar based systems, with a somewhat limited resolution of 1 m, due to the wavelength difference of optical light and microwaves as well as volume and mass limits when it comes to launching large mirrors and/or microwave antennas (refer to Fig. 17). This limited resolution however is largely offset by the feat that Radar satellites provide for observational capabilities at night and when a particular patch of land or sea is hidden beneath clouds. In addition radar satellites enable different applications than optical satellites and hence can largely complement humanity's picture of planet Earth.

- **KH-5: Argon (1962–1964; 6/12), m: 1274 kg**
- **KH-6: Lanyard (1963; 1/3), m: 1500 kg**
- **KH-7: Gambit (1963-1967; 36/38), m: 2000 kg, σ: 45 cm**
- **KH-8: Gambit (1966-1984; 57/60), m: 3000 kg, σ: 45 / 15 cm**
- **KH-9: Hexagon (1971-1986; 19/20), m: 11400 kg**
- **KH-10: Dorian/MOL (1963-1969; 0; 1 G$)**
- **KH-11: Kennan/Crystal (1976-1988; 9/10), m: 13500 kg**
- **KH-12: Improved Crystal (1990- ; 4/4), m: 19600 kg, life time: 10-12 years, σ: 10 cm**

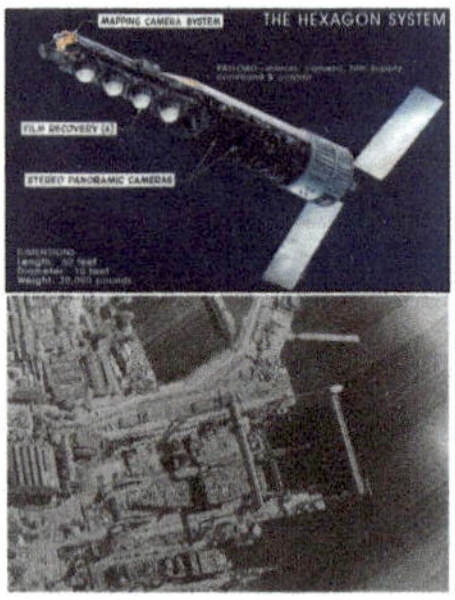

Fig. 10 A list of US spy satellite programmes along with two images depicting KH-9 hexagon, the last system, which relied on chemical film as well as a computer enhanced CCD-based KH-11 photo, showing the general layout of the Nikolaiev 444 shipyard in the Black Sea, with a Kiev-class aircraft carrier under construction. *Source* NRO and GWU (Images taken from; https://en.wikipedia.org/wiki/KH-9_Hexagon#/media/File:KH-9_HEXAGON_satellite.jpg and https://nsarchive2.gwu.edu/NSAEBB/NSAEBB13/)

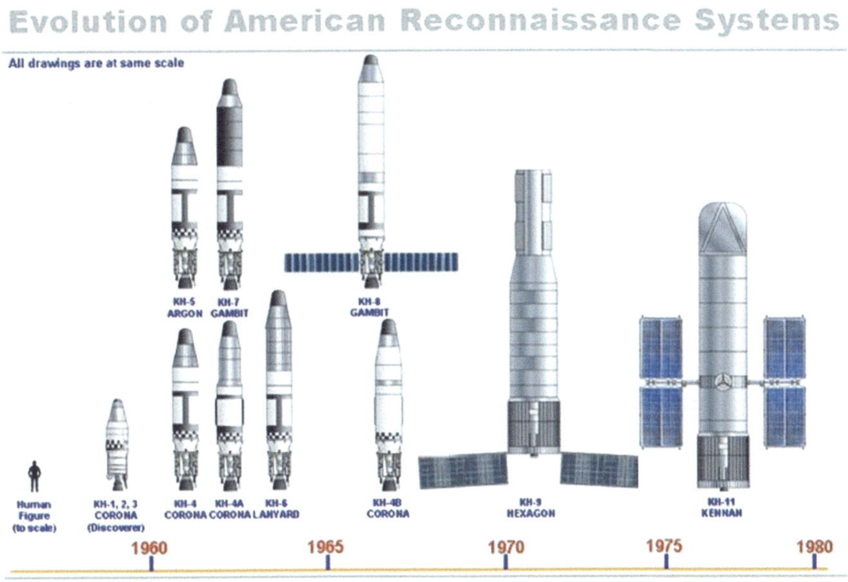

Fig. 11 Size comparison of US spy satellite systems. *Source* Giuseppe de Chiara (Image taken from: http://www.thespacereview.com/article/1809/2)

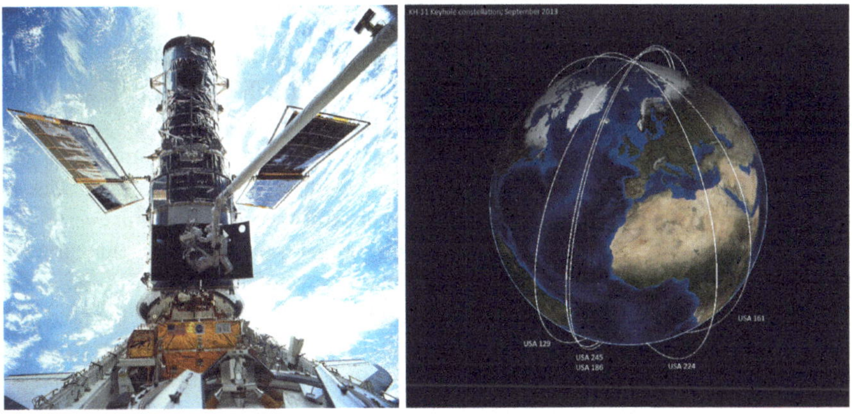

Fig. 12 Left image: Astronauts Steven L. Smith, and John M. Grunsfeld at work on the Hubble Space Telescope during the servicing mission 3A. *Source* NASA/ESA (Image taken from: https://www.spacetelescope.org/images/sts103_713_048/). Right image: KH-11 satellite constellation as of September 2013. *Source* Marcoaliaslama (Image taken from: https://en.wikipedia.org/wiki/KH-11_Kennen#/media/File:KH-11_Keyhole_constellation,_September_2013.png)

Figure 13 provides for a comparison how the same object will appear in space-based reconnaissance sensors, featuring different capabilities in terms of spatial resolution. Bearing in mind that a radar satellite will in general feature a

Fig. 13 A comparison how the same area of interest will look differently when observed by satellites with different capabilities in terms of spatial resolution

lower resolution than an optical satellite, one can say that the left image patch of Fig. 13 is attributed to the observation domain of radar satellites while the right image patch of Fig. 13 represents the capabilities of optical satellites. As such one could say that the bottom right image in the right patch represents the best observational capabilities of a modern radar satellite or of a commercial optical satellite of the first generation, such as IKONOS, while the left upper image in the right patch showcases the capabilities of the KH-11 and KH-12 satellite systems.

Today it is not only the United States that uses Earth Observation satellites; Russia, China, Japan, India, Europe and numerous other countries are relying on the continuous view on Earth. The reason for doing so ranges from pure commercial and civilian purposes to 100% military objectives with a floating dual-use area in between. And while modern peace-loving societies often tend to condemn everything that is related to military use, they shall not forget that over the last decades it were mostly the spy satellites that have tracked military movements, were instrumental of assessing the "bomber gap" and provided insights into the planning of the cold war adversaries. In the end, one may even argue that it was due to information from the spy satellites that the cold war remained cold and never turned into a hot war.

6 Private-Public-Partnerships for Space Reconnaissance Projects

Triggered by the success of spy satellites, the security dimension of space increased over time. Today, space is used for navigation, space reconnaissance (esp. ICBMs), control of own ICBMs and long range guided missiles, communication purposes, early warning, and more. However times have changed; the Cold War is history and so is the threat of nuclear annihilation and the confrontation of West vs. East, which provided reason for the incentive to run large space programmes like Apollo—using space as an arena to show ones scientific, technologic, economic and even societal strength to the adversary.

Still, national security remains one of the primary objectives of many countries and as Earth Observation has proven its worth for other application beyond reconnaissance, like urban planning, environmental monitoring and protection, tracking applications, etc., Dual-use is the buzzword of the hour. Europe's Copernicus programme is a master showcase for a Dual-Use Earth Observation programme, while GeoEye serves as a great example for an Earth Observation satellite built in a Private-Public-Partnership (PPP) fashion (refer to Fig. 14).

Launched in 2008, GeoEye-1 is a proponent of the most recent Earth Observation satellite generation, serving both the interests of civilian and military users. As such it features a very high resolution of 0.41 m and can perform observations on several bands and at large angles.

While resolution is always a key specification of every Earth Observation satellite one particular difference between the typical commercial/civilian and the military user is the maximum time that it may take until an acquired image is made available to the user and how quickly a specific area of interest can be revisited. Time is of essence—in particular for the military user. In a military campaign one wants to know as quickly as possible what adversaries (and own assets) are doing, how a campaign is progressing and how things are developing ("change detection"). Therefore image transmission and processing needs to be fast and several satellites need to be able to provide for a frequent observation of the area of interest. Consequently companies like the Satellite Image Corporation have not only one satellite like GeoEye-1 in their portfolio but several, with the latest (WorldView-4)

GeoEye-1:

- **Launch: 06.09.2008, Delta II, Vandenberg**
- **Operator: GeoEye (USA), Merger: Space Imaging + OrbView**
- **Design: General Dynamics Advanced Information Systems / ITT Corporation**
- **Mass: 1955 kg; 4,35 x 2,7 m**
- **Costs: 502 M$ (207 M$ via U.S. National Geospatial Agency NextView Programme)**
- **Orbit: SSO, 81°, 681 km, 15 orbits/day**
- **Life time: 7-15 years, 700.000 km²/day**
- **4 bands, σ: 0,41 m*, 15 x 15 km**

Fig. 14 Key features of the GeoEye-1 satellite. *Source* DigitalGlobe (Image taken from: https://www.satimagingcorp.com/satellite-sensors/geoeye-1/)

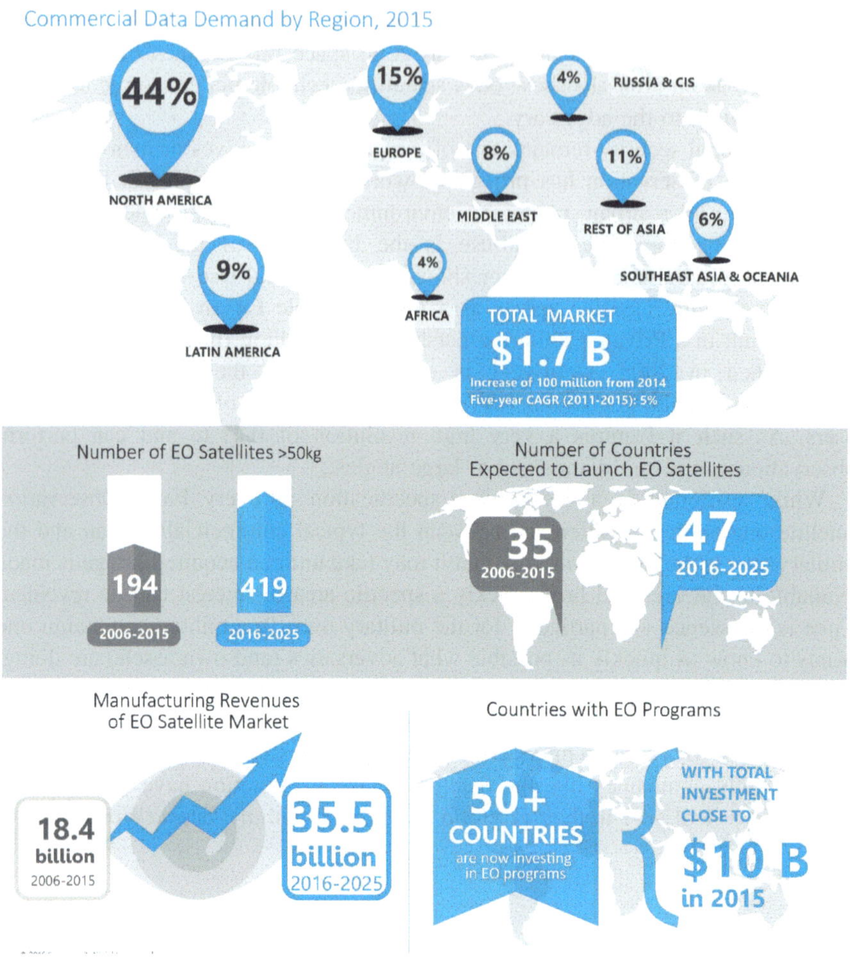

Fig. 15 Current and forecasted commercial data demand in the earth observation market for different regions *Source* Euroconsult

offering a resolution of 0.31 m (close to the right upper image in the right patch of Fig. 13).[15]

As the applications and services in the space and security domain have increased so have the number of players. Today, Earth Observation is one of the most vibrant application domains in space, serving both civilian and military users often at the same time (refer to Fig. 15).

[15]Satellite Imaging Corporation [15].

7 Small Is Beautiful? Small Sats and Earth Observation

As picturesque as a high resolution image of a particular place on Earth may be, several applications and services, like harvest forecasting, deforestation monitoring, EO data collection for insurance purposes, urban planning do not demand Earth observation data in the cm-range. For such applications, resolutions in the order of 2.5–5 meters are sufficient (refer to Fig. 13 left patch of images). In addition, these resolutions are also the ones of choice for "Change Detection" approaches employed by military users (refer to the paragraph below Fig. 14)—best realised by a mix of very few high resolution satellites and a large constellation of low resolution satellites. While the high resolution satellite will observe the areas of interest once every few days or weeks, a fleet of low resolution satellites will provide for a very frequent and wide coverage, making it easy to detect any changes. If these exceed a certain threshold and hence require the observer to take a closer look, a higher resolution satellite is tasked to make an up-to-date observation of the area of interest and the complete change detection observation cycle starts again.

The whole procedure is similar to the optical compression algorithm employed in DVDs or Blu-rays, where the requirement of having 25 frames per second while reducing the data to be stored to the greatest extent has sparked the idea of using key and delta frames. While the key frame represents a full picture and hence takes a huge amount of storage data, the delta frames in between contain only the data set, which has changed with respect to the key frame (e.g. the moving mouth when two people are standing still in front of s static background while talking with each other). Most of the compression algorithms like MPEG use one key frame every 5–10 s and hence save a lot of storage capacity.[16] Obviously the similarity to the change detection is that the key frame is like the high resolution satellite image, while the delta frames can be compared to the low resolution and frequent satellite observations. Under this picture a company like Planet, with its 3–5 m resolution nano satellite Earth Observation constellation provides for the delta frame functionality within the change detection algorithm (Fig. 16).

What sounds like an easy task, is however quite difficult. Nano satellites like CubeSats are very small; the smallest unit—1U—is as "big" as 10 × 10 × 10 cm and allows for a maximum mass of 1.33 kg per unit. Combining multiple units like 2U, 3U, 6U, etc. is possible. Trying to implement a telescope into such a small volume is an arduous task but absolutely necessary to obtain a 3–5 m resolution. In the case of Planet, this meant that one had to design a 3U CubeSat in a way that it was virtually built around the telescope. The physics behind the optical capabilities of an Earth Observation satellite, as discussed within Fig. 17 demonstrates the difficulties if one wants to achieve a spatial resolution in the order of a few centimeters.

Obviously the aperture of the optical sensor as well as the distance to the object are of essence to achieve really high resolution. Building upon the formula depicted

[16]Page [16].

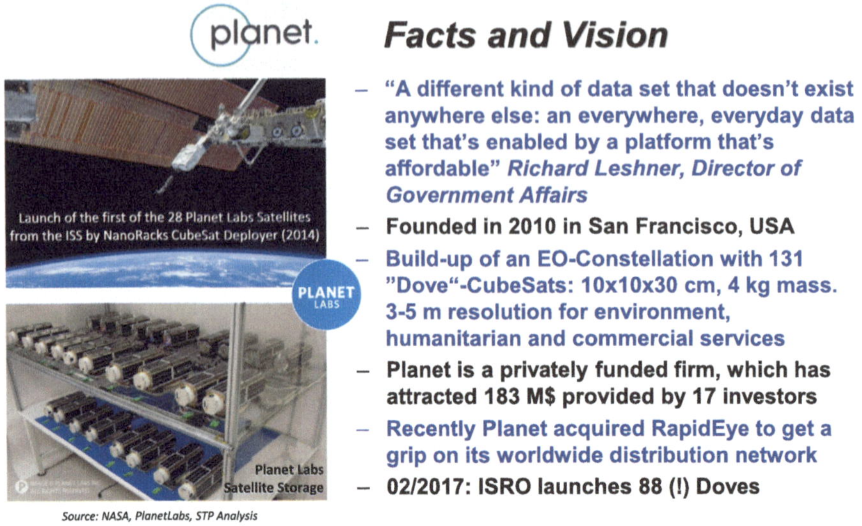

Fig. 16 Planet's CubeSat earth observation constellation between facts and vision

on top of Fig. 17 explains how satellites like KH-11 can achieve high-end reconnaissance features; only by being placed in a low Earth orbit and using a mirror with a diameter of 2.34 m it is able to achieve a spatial resolution of a few centimetres.

Acknowledging that the HST is in essence nothing else but a KH-11 looking in the other direction, we can use the same methodology for an astronomical nano satellite, like the Uni-BRITE/BRITE-Austria, which is part of a constellation to

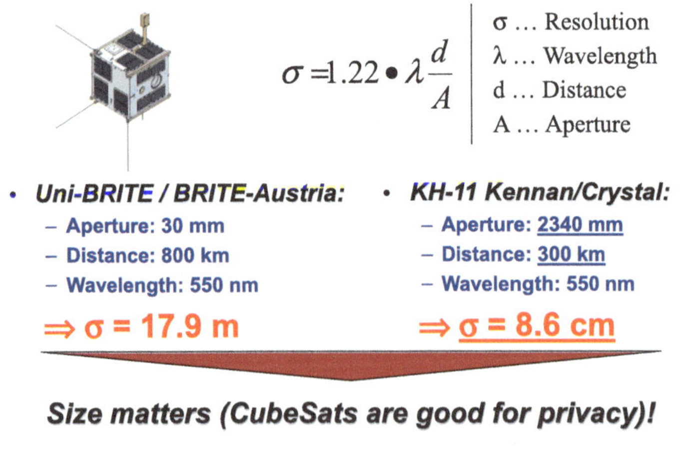

Fig. 17 The physics that drives the optical resolution that can be achieved by an earth observation satellite

measure brightness variations of very bright stars, such as can be found within the constellation of Orion. What if we follow the KH-11/HST swap and turn Uni-BRITE/BRITE-Austria around, using its optical telescope to observe the Earth? What kind of resolution would we achieve? The left side of Fig. 17 provides the answer; as the aperture of the telescope is limited to 30 mm and the distance is 800 km the resolution is mediocre; 18 m is what can be achieved with such a system.

While this may be disappointing at first, a closer look reveals the positive side to it. Acknowledging that more and more small satellites like CubeSats are launched into space to perform Earth Observation tasks, one could think that George Orwell's "Big Brother" vision is about to be realised by numerous eyes in the sky constantly tracking everyone's whereabouts. Figure 17 however tells us that this is not true, simply because of physics summarised by one statement: "size matters". If a telescope is too small, the resolution will be mediocre. Therefore CubeSats should be seen as a great Earth Observation asset as they allow us to constantly observe and monitor our Earth while providing us with sufficient space to protect the privacy of the people that are living on this planet.

Acknowledgments The author would like to acknowledge Prof. René Laufer, Baylor University, for his insights and comments, which became very valuable contributions to this article.

References

1. Huggel C, Kääb A, Salzmann N (2006) Evaluation of Qucikbird and IKONOS imagery for assessment of high-mountain hazards. In: EARSeL eProceedings 5, 1/2006 http://www.eproceedings.org/static/vol05_1/05_1_huggel1.pdf. Accessed 9 Jan 2018
2. Satellite Imaging Corporation (2017) QuickBird satellite sensor. https://www.satimagingcorp.com/satellite-sensors/quickbird/. Accessed 08 Jan 2018
3. Baucom D (1997) Missile defense milestones 1944–1997. https://web.archive.org/web/20160306023546/fas.org/spp/starwars/program/milestone.htm. Accessed 8 Jan 2018
4. U.S Strategic Command (2017) History. http://www.stratcom.mil/About/History/. Accessed 08 Jan 2018
5. U.S Strategic Command (2017) Command snap shot. http://www.stratcom.mil/About/Command-Snapshot/. Accessed 8 Jan 2018
6. Wikimedia Commons (2010) Seal of the United States strategic command. https://commons.wikimedia.org/wiki/File:Seal_of_the_United_States_Strategic_Command.svg. Accessed 8 Jan 2018
7. Encyclopaedia Britannica (2018) Missile gap. https://www.britannica.com/topic/missile-gap. Accessed 8 Jan 2018
8. Military Factory (2017) Lockheed Martin U-2 dragon lady high-altitude reconnaissance aircraft/spyplane (United States). https://www.militaryfactory.com/aircraft/detail.asp?aircraft_id=51. Accessed 8 Jan 2018
9. CIA (1992) The CIA and overhead reconnaissance—the U-2 and OXCART programs, 1954–1974. https://www.cia.gov/library/readingroom/docs/DOC_0000190094.pdf. Accessed 9 Jan 2018
10. CIA (1992) The CIA and overhead Reconnaissance—the U-2 and OXCART Programs, 1954–1974. https://www.cia.gov/library/readingroom/docs/DOC_0000190094.pdf. Accessed 9 Jan 2018

11. Breaking Defense (2015) Air force, riding budget boost, warns on sequester; U-2 Is BACK! https://breakingdefense.com/2015/02/air-force-riding-budget-boost-warns-on-sequester-u-2-is-back/. Accessed 9 Jan 2018
12. Ruffner K, CIA, Center for the Study of Intelligence (U.S.), History Staff (1995) Corona: America's first satellite program. http://www.worldcat.org/title/corona-americas-first-satellite-program/oclc/42006243. Accessed 9 Jan 2018
13. Brown S (1996) America's first eyes in Space. In: Popular science. Feb 1996, p 46–47
14. Chun R (2015) How to recreate Howard Hughes' legendary movie screening room. Wired. https://www.wired.com/2015/12/howard-hughess-screening-room/. Accessed 10 Jan 2018
15. Satellite Imaging Corporation (2017) WorldView-4 satellite sensor. https://www.satimagingcorp.com/satellite-sensors/geoeye-2/. Accessed 10 Jan 2018
16. Page N (2000) Nicky pages' digital solutions. Keyframes & delta frames explained. http://nickyguides.digital-digest.com/keyframes.htm. Accessed 10 Jan 2018

Norbert Frischauf has studied Technical Physics at the Technical University in Vienna and Electrical Engineering at the Technical University in Graz (Austria). Following his specialization on high energy physics he moved to CERN in Geneva (Switzerland) to work on two particle detectors in the DELPHI experiment of the Large Electron Positron Collider (LEP).

After having spent some years at CERN, his professional carrier led him to the European Science and Technology Centre (ESTEC) of the European Space Agency (ESA) in Noordwijk (Netherland), where we worked as future studies systems engineer and expert for emerging technologies.

From 1999 to 2006, Norbert was engaged with consultancy work for Booz Allen Hamilton, mostly focusing on aerospace, telecoms and hi-tech.

After a professional detour into management with the start-up QASAR in Vienna, Norbert returned to the Netherlands in the beginning of 2009, where he worked for three years as a scientific officer at the EC Joint Research Centre—Institute for Energy and Transport in the action for "Hydrogen Safety in Storage and Transport", being responsible for scientific aspects of high pressure hydrogen storage activities and further technical developments.

As of 2012 Norbert is a partner at SpaceTec Partners a unique boutique consultancy providing strategy and technology consulting, communication activities and interdisciplinary project management mainly for the European Commission, ESA and leading industries. Being the "R&D and hi-tech partner", Norbert's role within the company involve: Strategy development, Science and technology competence, Experimental research, Networking, as well as Authoring and presenting.

General Overview—Economic Aspects

Alessandra Vernile

Space offers an important insight to understand the Earth and to improve our lives. Satellite Earth Observation (EO) can benefit many areas of society, including environment and resources management, agriculture, food security, transport, just to mention few of them.

The importance of EO is highly relevant for the improvement of societal needs. This relevance has been also recognised by policymakers that are now promoting the use of space-based EO data and products through new technology policies. These policies are able to take into consideration the benefits that the space sector, and in particular the EO satellite-based services, would give to the society. In particular, the policymakers' focus is based on the motivations that would help the investments in space-based satellite technology for EO:

- Space asset continuity
- Data and product accessibility
- Performance
- Affordability.[1]

According to the "Space Report" 2016 edited by the Space Foundation,[2] the global space industry grew in 2015 reaching the amount of $323 billion. A relevant role has been played by the revenues from commercial sectors that are representing a three-quarter of the global economic activity in space. The commercial space

[1]Benefit Assessment of the Application of Satellite Earth Observation for Society and Policy: Assessing the Socioeconomic Impacts of the Development of Downstream Space-Based Earth Observation Applications, Murielle Lafaye, in Onoda and Young [1]

[2]The Space Report 2016. Overview. The Space Foundation. June 28th 2016. https://www.spacefoundation.org/sites/default/files/downloads/The_Space_Report_2016_OVERVIEW.pdf.

A. Vernile (✉)
Eurisy, Paris, France
e-mail: alessandra.vernile@eurisy.org

C. Brünner et al. (eds.), *Satellite-Based Earth Observation*,
https://doi.org/10.1007/978-3-319-74805-4_6

products and services, which includes among the others EO, constitute the biggest part of the sector, reaching in 2015 $126.33 billion.

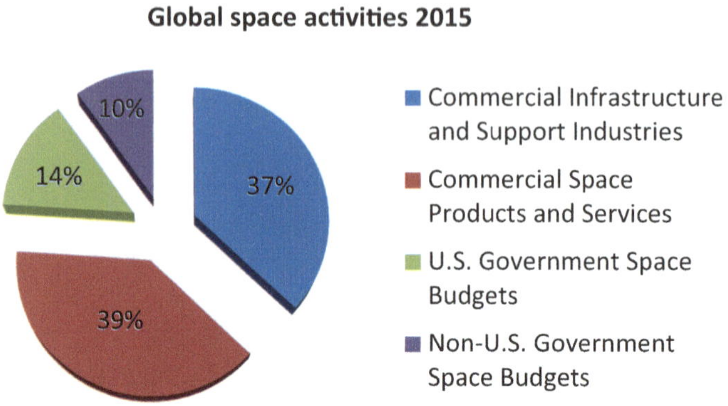

Looking at the 2016 "State of the Satellite Industry Report" edited by the Satellite Industry Association (SIA) and Bryce Space and Technology, the satellite revenues remained the largest industry segment, powered by the EO services and communication satellites. In particular, the 62% of the space market is occupied by satellite services and, in particular, of the 1381 satellites launched in 2015 the 14% of them is dedicated to EO, growing by the 10% compared to the previous years.[3] This growth can be ascribed to the growing number of newly established remote sensing companies, mostly start-ups and SMEs and from acquired satellites and Mergers and Acquisition, and new technologies. These new entrants are raising capital easily, compared to the already well-established traditional companies, thanks to the support of the public sector and the relevant role played by business angels, venture capitalists, and other financial funds. The entrants of new actors in the space sector and in particular in the EO segment, guarantee a technological development of new satellites as well as the initial deployment of satellite constellation for EO.

The EO data exploitation market, which includes sales and value-added services, benefits from the annual growth of all the space exploitation markets. Today the need for EO data and services is living a new era, due also to the central role of the big data and the new commercial activities, conducted by start-ups and/or Small and Medium Enterprises (SMEs). These factors make the EO market a highly developing market for the next years.

Both downstream and upstream sectors are growing fast. In particular the EO downstream market, that includes companies developing added-value services based on EO data, is becoming very mature and is providing a series of opportunities in different sectors. This is true also because of the recent technological progress that

[3]State of the Satellite Industry Report 2016. Satellite Industry Association. March 2016. http://www.sia.org/wp-content/uploads/2017/03/SSIR-2016-update.pdf.

enables companies to present innovative solutions for EO creating a major competition within the market and for their potential customers.

The downstream market for EO applications can be divided into five areas:

1. Natural resource management
2. Defense and security
3. Land Monitoring
4. Oceanography
5. Meteorology

1. Natural resource management

Natural resources include biological resources, minerals, forests, farmland, water, and oceans. The monitoring of these resources would result to be complicated without the support of EO satellites that are able to observe wide areas of the planet.

Space and space-based technologies offer supportive solutions to the monitoring of natural resources, in particular in the frame of international activities, such as the UN Sustainable Development Goals (UN SDGs) and the UN Agenda 2030.[4] Remote sensing data help to have a complete image of the Earth thanks to the periodic observations and the inputs into decision-making processes, assuming an important role in the achievement of the goals and the development of society. For this reason, it is necessary to analyse the subcategories identified in the light of the UN SDGs.

Forestry

Forests have a crucial role in reducing the risk of natural disasters and help in mitigating climate changing. The involvement of remote sensing is important for reconnaissance mapping, whose aim is to meet the needs of forestry/environment agencies for scientific scopes.[5] Commercial forestry reveals to be important for resource management agencies to collect information and to offer support to conservation authorities involved in monitoring the quantity, quality, health and diversity of forests.[6] Different are the activities conducted at institutional and private level. It is important to mention the European Union Programme Copernicus and its operating satellites Sentinels.[7] In particular, the Sentinel missions support climate

[4]United Nations Sustainable Development Goals. https://sustainabledevelopment.un.org/sdgs.

[5]Mapping the world's forests. European Space Agency. http://www.esa.int/Our_Activities/Observing_the_Earth/Mapping_the_world_s_forests.

[6]United Nations SDG No. 15. https://sustainabledevelopment.un.org/sdgs.

[7]Copernicus, previously known as GMES (Global Monitoring for Environment and Security), is a European Union Programme aimed at developing information services based on satellite Earth Observation and in situ non-space data. Copernicus is coordinated and managed by the European Commission and implemented by partnerships with the Member States, the European Space Agency (ESA), the European Organisation for the Exploitation of Meteorological Satellites (EUMETSAT), the European Centre for Medium-Range Weather Forecasts (ECMWF), other EU agencies and Mercator Océan. Copernicus operates through six thematic streams of Copernicus services: Atmosphere, Marine, Land, Climate, Emergency, and Security. The main users of Copernicus

change studies providing data that can be used to supplement time-series acquisitions and providing additional value to the data acquired by other European and international EO programmes supporting response and management of political issues. Is Sentinel-1 the satellite that provides data on forest fire scar mapping, used to map the history and lifecycle of forests considering the carbon content. As private initiatives, it is relevant to mention the EO activities of the U.S.-based start-up Planet, that focus on forest health monitoring, illegal logging tracking, and operations planning. Through this activity, Planet is trying to revolutionise the access to forestry information to make forestry operations more accessible. Furthermore, Planet's satellite constellations have proven to be an asset for the REDD+ initiative to monitor the MRV, providing an EO-based monitoring programme to help the customers to assess and quantify the change in forestry.[8]

Agriculture

Satellites are used for agricultural purposes in different ways, and in particular for crop yields. The importance of applying satellites to agriculture can help in providing a precise picture of the areas to be cultivated, helping the differentiation between crop types and to assess their health and maturity. This information can help the improvement of the market and provide an early warning crop failure or famine. The information gathered by EO satellites are often combined with Geographical Information Systems (GIS) to allow a more intensive and efficient use of territory and better cultivation practices. Keeping in mind that the 37% of Earth's land is employed for agricultural purposes, the implementation of an EO system for agriculture is proven to high impact on the society and on the future development of it.[9] Also in this field, Copernicus' Sentinels are proving to be important to offer the society a comprehensive image of the Earth and to facilitate policy-makers in their task of setting up profitable policies for agricultural improvement.[10]

Water management

The management of water resources is important due to the number of countries in the world that cannot have access to water and to understand the hydrological cycle and the abundance or scarcity of this resource.[11] For this reason both UN and the World Bank recognising the role of EO for water resources, set up an operational framework to meet the needs of users and policymakers.[12]

services are policymakers and public authorities who need the information to develop environmental legislation and policies or to take critical decisions in the event of an emergency, such as a natural disaster or a humanitarian crisis. http://www.copernicus.eu/main/copernicus-brief.

[8]Planet. Markets overview. https://www.planet.com/markets/redd/.

[9]https://earth.esa.int/web/guest/earth-topics/agriculture.

[10]Sentinel-2 and Landsat-8 are highly involved in the mapping of cultivable areas and for the assessment of land health. http://www.esa-sen2agri.org/operational-system/eo-data/.

[11]https://earth.esa.int/web/guest/earth-topics/water/water-management.

[12]Garcia et al. [2]. https://openknowledge.worldbank.org/bitstream/handle/10986/22952/9781464804755.pdf.

2. Defence and Security

EO for Defence and Security represents a relevant slice of the EO industry, from an economic and political angle. According to the SIA Report on State of the Industry 2016, the percentage of the satellites sent into space for defence purposes is about 8%. The demand of national government supply and demand and the new commercial solutions are increasing since a relevant number of countries are using their EO defence capacity in different fields.

According to Euroconsult,[13] in the timeframe 2006–2015 only six countries applied EO defence capacity and, according to a report from Business Wire Company,[14] in the future, this number will probably remain static. A wind of change can be expected by the private actors, new entrants in the market with the aim of cutting down the costs of space solutions and services. An increasing role is played by dual-use satellites, especially for governmental uses. It is important to mention in this frame the COSMO Sky-Med satellite constellation developed by the Italian Space Agency and the Italian Ministry of Defence.[15]

3. Land Monitoring

To guarantee the monitoring of the lands, and the vegetation state or water cycle, different programmes have set up to offer a picture of the Earth and to give information on land cover, land use, cover-use changes, vegetation state or water cycle. In this regard, may be useful to recall again Copernicus and in particular the Copernicus Land Monitoring Service,[16] whose major components are global, pan-European and local. At European level, it has been recognized the importance of land monitoring, due to the highest share of land on the European territory used for different purposes.

Many drivers show the importance of having an EO support for land use, in Europe and everywhere else: land monitoring is important due to the increasing

[13]Earth Observation: Defense and Security. A Euroconsult Executive Report. 2016 Edition. Euroconsult. http://www.euroconsult-ec.com/research/earth-observation-defense-security-2016-brochure.pdf.

[14]Earth Observation: Defense and Security Market Report 2016—Analysis, Technologies and Forecasts—Key Vendors: Airbus Defence and Space, DigitalGlobe, ImageSat - Research and Markets. Business Wire.
http://www.businesswire.com/news/home/20160420006225/en/Earth-Observation-Defense-Security-Market-Report-2016.

[15]Cosmo-Skymed. http://www.e-geos.it/cosmo-skymed.html.

[16]Copernicus FactSheet. Spacetec Partners. http://www.spacetecpartners.eu/copernicus-factsheets/Services/03_Land_V2.pdf.

need of space per person and for economic activity, the increased mobility and the growth of transport infrastructure.[17]

4. Oceanography

As well as land monitoring is necessary also the study and the observation of the oceans to implement the studies on climate changing, assess the health of the oceans and help the developing of the blue economy regarding the fisheries and the conservation of marine species. National space agencies around the world set up programmes to offer a complete coverage of the water surfaces principally for scientific purposes, an example can be the Franco-American mission "Surface Water and Ocean Topography" (SWOT), projected to be launched by 2021.[18] The protection of the marine environment and the contribution for development is something that also UN took in serious consideration in the Agenda 2030 and the SDGs. The UN promotes a good governance and hoped for sustainable marine-based activities achievable thanks to the support of space-based technologies and to EO satellites.[19] Also, the European Copernicus and its Sentinels are involved in oceanography, especially thanks to the Copernicus Marine Environment Monitoring Service (CMEMS). The CMEMS provides information on the ocean and marine ecosystems both globally and European and contributing to the marine safety and the sustainable management of marine resources.[20]

5. Meteorology

EO contributes also to metereological purposes: the metereological missions served by EO satellites, allow the end users to have cost-efficient data, images, and products for commercial and scientific aims. This permits to cover the Earth for weather and climate monitoring. There are different geostationary satellites dedicated to meteorology as well as the involvement of national agencies. In this frame, it must be mentioned EUMETSAT and their detailed systems for EO provided by Meteosat and MetOp, in collaboration with the Member States of the organisation and ESA. The relevance of studying the meteorological phenomenon from space is important for the intertwining with the climate monitoring, as today's challenges for the impact on societal needs and the development of the population on Earth and the economic features.[21]

---

[17]Copernicus. http://land.copernicus.eu/.

[18]CNES and NASA Sign Statement on Mars and Oceanography. Mission pour la Science et la Technologie de l'Ambassade de France aux Etats-Unis. June 20th 2017. https://www.france-science.org/CNES-and-NASA-Sign-Statement-on.html.

[19]United Nations Sustainable Development Goals. https://sustainabledevelopment.un.org/topics/oceanandseas.

[20]Copernicus. http://copernicus.eu/main/marine-monitoring.

[21]What we do, Eumetsat. https://www.eumetsat.int/website/home/AboutUs/WhatWeDo/MonitoringWeather/index.html.

1 Revenues Influential Factors

A number of factors that affect EO industry have been identified. EO providers are concentrating their efforts in the primary market. These primary markets are the best chance to increase the revenues in the market, especially on the services' side. The expected growth relies also on the development of new technologies, such as small satellites, and on new customers for new applications.

The growth of EO market involves creates high expectations. These expectations picture a new level of data service that offers an increased demand for EO imagery. This demand is a combination of demands coming from different sectors (IT, mining, agriculture, oil and gas, financial companies, governments, fisheries, etc.) and that results of a new role of data in the decision-making process.

The development of new satellite navigation techniques, also modeled on the Google-Earth-like model, is impacting the market bringing changes in the market for EO products. According to some analysts, these changes will significantly revolutionise the market in the medium term, contributing to the next development of the market.[22]

The impact on EO depends on three major group of factors that can influence Earth Observation sector:

1. Technological factors
2. Political and regulatory factors
3. Social factors.

Social, demographic, economic and political factors are impacting on EO revenues. In addition to this the rise of private actors and the development of new services and technologies for some of the sectors mentioned earlier, are extending the demand more and are making the EO space sector an inclusive one, also for non-space actors.

Looking at the technological trends, it is important to recognise the increasing amount of new satellites launched for EO purposes and the development of payload technologies and new sensors. In addition to this, with the rise of private actors, the projects mega-constellations of microsatellites for EO and for telecommunications are regaining strength. The most relevant example is the mega-constellation for EO of Planet, formerly Planet Labs. The so-called "doves" together form a satellite constellation that provides a complete image of the Earth at a 3–5 m optical resolution and open access data. The set-up of this constellation of satellites, embraces the idea of a democratisation of space, which includes the concepts of affordability and accessibility of data and technologies for everyone. This democratisation impacts on the increment of the revenues for the space sector, because of the development of new elements in the satellite supply chain and new customers.

---

[22]Harebottle, A. "Earth Observation to Capture the Mainstream Market". Via Satellite, May–June 2017. http://interactive.satellitetoday.com/via/may-june-2017/earth-observation-to-capture-the-mainstream-market/.

Planet sells both to institutional and private customers, introducing a low-cost service that provides information for climate monitoring, crop yield prediction, urban planning and disaster response.[23]

Besides this, advances in geographical information systems and digital mapping provide the necessary support to locally based services and new services that help the continuity of data. The result of this is the impact on cost and reliability of these services and the convergence and integration of satellite capabilities with terrestrial ones, this implies new target and new customers for the entire market. The objective is the demonstration of the benefits of EO technology and services to potential users.

Due to new directives environmental monitoring[24] and the related policies in fishery policies, etc., the demand for EO services increased. In addition to this, and in coincidence with the definition of the UN Agenda 2030 and the Sustainable Development Goals (SDGs), supply and demand for EO technologies and satellite-based services had also increased due to new measures for regional development or humanitarian aid both for developing and developed countries, to help with a sustainable growth of society.

In general, the space sector became a sector where private supply and public demand can meet even if we are assisting to a standardisation of the market who is embracing new standards. This is particularly true with the participation of private actors in space activities since one of their features is the standardisation for a reliability of the product and a low-cost approach. If in the future these standards will be applied to the whole space sector, the EO market will be a more encouraging environment for different actors, also non-space actors, and will increase the production of interoperable equipment and mass production, as it is happening with the space transportation market.[25]

EO asks also for a better environmental law, especially in these days since EO can give an important contribution to earth monitoring. Indeed, satellite data can be used in monitoring international agreements as it happened already in the past with the Globwetland projects, funded by ESA, support the Ramsar Convention on the conservation of wetlands and waterfowl through the use of satellite data. Globwetland I and Globwetland II demonstrated the potential of satellite data to

[23]50 Companies That May Be the Next Start-Up Unicorns, New York Times, Aug. 23rd, 2015, http://bits.blogs.nytimes.com/2015/08/23/here-are-the-companies-that-may-be-the-next-50-start-up-unicorns/. See also, 50 Future Unicorns predicted by CB Insights Company Mosaic, CBInsights, Aug. 23, 2015. https://www.cbinsights.com/research-future-unicorn-companies.

[24]Adoption of the Paris Agreement. United Nations Framework Convention on Climate Change. Conference of the Parties Twenty-first session Paris, 30 November to 11 December 2015 Agenda item 4(b) Durban Platform for Enhanced Action (decision 1/CP.17) Adoption of a protocol, another legal instrument, or an agreed outcome with legal force under the Convention applicable to all Parties. FCCC/CP/2015/L.9. December 12th 2015.

[25]Business in Earth Observation. "An overview of market development and emerging applications". EOVox, 2016. http://earsc.org/file_download/43/Business+in+Earth+Observation+eoVOX 080508.pdf.

monitor wetlands.[26] These projects and initiatives, such as the Treaty Enforcement Services using EO, help in assist in development serving as a basis from where the use of satellite can be extended to verify the development and the standards to which satellite data has to be stick to not violate international agreements. For this reason, it is important to development also a political framework in which environmental laws can be included.

The data provided by EO satellites can help in increasing the relative measures to adhere to international treaties and agreements. For this reason, it is important to also develop a dedicated economic framework for EO sector and to guarantee interoperability and sustainability of this market.[27]

2 New Challenges Ahead

The challenges ahead for the EO market can be of three kinds and are perfectly in line with the actual trends that guide the market now:

1. Geopolitical changes
2. Environmental commitments
3. Energy market changes

These new challenges are intertwined. The focus of the attention today has been moved to the global political challenges and security concerns. These changes, together with the new environmental commitments taken by the international community with the COP21 and 22, that involve also energy market changes, can have a relevant impact on the EO market in the future. For this reasons, it is important to examine the space opportunities in the international evolving context identifying the contribution that the earth observation technologies can offer for instance in the tracking of migrant routes, for the observation and mission to study the atmosphere, the level of the sea and the emission of greenhouse gases and so on.[28]

[26]European Space Policy Institute (ESPI). Satellite Data to Monitor International Agreements. ESPI Brief No. 14 June 2017. Retrieved: https://www.espi.or.at/images/ESPI_Brief_14.pdf.

[27]ESPI, *in ibid.*

[28]Kuan-Wei Chen. Satellite Usages and the Development in Space Law and Policy: Challenges and Opportunities in the Global Governance Study. PowerPoint Presentation. Centre for Research in Air and Space Law McGill University Montreal, Canada. Asia-Pacific Space Cooperation Organization SPACE LAW and POLICY FORUM 2017 10–11 July 2017 Harbin, China.
https://www.mcgill.ca/iasl/files/iasl/satellite_usages_and_the_development_in_law_and_policy_global_space_governance-perspectives.pdf.

3 The EO European Space Market

In the international context, it is important to take a look at the European context and the EO market development. First of all, it is important to delineate the context in which the EO market is evolving.

The European industrial complex is a skilled and motivated technology industry that has been able to achieve a relevant role on commercial markets, contributing to the growth of the European economy, but it is presented as a fragmented market. The relevant role of Europe in the space sector is related to the space business and launch service market, expressed also in the EC European Space Strategy for Europe.

Under the lens of EU Member States, there is the Copernicus programme, as outlined in the EU Space Strategy for Europe, issued on October 2016.[29] The Copernicus programme is a user-driven programme which provides operational services such as atmosphere monitoring, marine environment and land monitoring, climate changing, emergency management and security. These services are provided to EU, national, regional institutions and private entities. Copernicus aims at filling the gap in the European EO capacities. Data is provided from space infrastructures together with the support of the Member States. ESA is in charge of the operational part of the mission, thanks to the Sentinels. Copernicus introduces the free data policy, which is important for a better development of the service, for the development commitments taken at a global level and to help the European space market to evolve. For this reason, Copernicus requires a high level of continuity in service and data and to provide the adequate funding for development of the operational EO missions and services. The final phase of Copernicus assures the development of the downstream sector of the European space market and would guarantee a long-term political and financial commitments for Europe.[30] The EU Space Strategy for Europe takes into consideration the EO data market in relation to Copernicus looking at the space data, improving the access and exploitation of them enabling a cross-sectoral data exchange.

In addition to this, Europe should open more towards private entities, whether is start-ups or SMEs. Keeping in mind that today NewSpace is able to highlight the crucial role played by the space assets to meet population needs, this creates also the right framework to guarantee the EO market to grow and to help to improve the economic growth and stimulate the job market. This can be done, as also outlined in the Space Strategy, through the implementation of PPP schemes and P2P solutions that would guarantee a facilitation for the downstream sector.

[29]Communication from the Commission to the European Parliament, the Council, the European Economic and Social Committee and the Committee of the Regions. Space Strategy for Europe. Brussels, 26.10.2016 COM (2016) 705 final. https://ec.europa.eu/transparency/regdoc/rep/1/2016/EN/COM-2016-705-F1-EN-MAIN.PDF.

[30]Reillon, Vincent. Securing the Copernicus programme: Why EU earth observation matters. European Parliament Think Tank. April 6th 2017. http://www.europarl.europa.eu/thinktank/it/document.html?reference=EPRS_BRI(2017)599407.

References

1. Onoda M, Young OR (eds) (2017) Satellite earth observations and their impact on society and policy. Springer, ISBN 978-981-10-3713-9
2. Garcia LE, Rodriguez DJ, Wijnen M, Pakulski I (eds) (2016) Earth observation for water resources management. Current use and future opportunities for the water sector. ISBN (electronic): 978-1-4648-0476-2. World Bank Group. https://openknowledge.worldbank.org/bitstream/handle/10986/22952/9781464804755.pdf

Alessandra Vernile is a project officer at Eurisy. From 2016 to 2017, she served as Resident Fellow at European Space Policy Institute (ESPI) with a fellowship from the Italian Space Agency (ASI) and the Italian Society for International Organization. Previously she worked as Intern at Strategy Department at European Space Agency HQ in Paris and as Event Manager at NATO Defense College Foundation in Rome. MA in International Relations at LUMSA University, in Rome. In 2014 she attended a Master in Economic Security, Geopolitics and Intelligence at SIOI and in 2015 she got as fellowship-student for the Master in Institutions and Space Policy at SIOI, in Rome.

Robocopter

Wolfgang Schinagl

1 Introduction

The Symposium, "Trends und Herausforderungen satellitengestützter Erdbeobachtung für Wirtschaft und Gesellschaft" was organised by the "Kompetenzzentrum für Weltraumrecht und Weltraumpolitik" located at the Karl Franzens University of Graz, Austria, headed by Em. Univ. Prof. Dr. Christian Brünner (Subpoint ECSL/NPOC Austria) on 1st of June 2017 in the Europa hall of the Styria Economic Chamber in Graz.

One major argument for the collaboration with Styria Economic Chamber was the economic impact of satellite-based earth observation in multiple industrial branches from agriculture to environment-technologies. The province of Styria is economically driven by research and development. Based on the latest surveys of Statistics Austria and Joanneum Research for 2015, the Austrian province of Styria has reached the highest R&D quota to 5.16% of its gross regional product (GRP), which is the first place in the ranking of the nine provinces of Austria and is the highest R&D quota of all 276 EU regions. The Graz University of Technology has a long history in satellite and space technology—e.g. Em. Univ. Prof. DDr. Willibald Riedler was responsible for arranging for Ing. Franz Viehböck to be the first Austrian cosmonaut to the space station Mir in 1991 for just over a week. Today the full spectrum of earth observation technology is a huge potential for new insights which will lead to powerful business applications. In this respect, popular quadrocopters, multicopters and the upcoming robocopters play an important role in the business world of the near future. Therefore, the author of this paper aims to show the benefits, potentials, restrictions and dangers of the next level multicopters: the robocopters.

W. Schinagl (✉)
WKO Stmk, Technische Infrastruktur, Digital Content Research & Development Center (DCRDC), Graz, Austria
e-mail: wolfgang.schinagl@wkstmk.at

C. Brünner et al. (eds.), *Satellite-Based Earth Observation*,
https://doi.org/10.1007/978-3-319-74805-4_7

2 The Noun: Robocopter

The word Robocopter is a neologism which is created by two compound nouns: "robot" and "helicopter". The word "robot" has its origin in the Czech language from the word "robota" which means "forced labour" and "slave labour". The word robot was popularised by the Czech author Karel Čapek and his science fiction play: "Rossum's Universal Robots" (1920). The etymology of the word "helicopter" is quite interesting, because it is itself a neologism, which was already used in 1887, when the Parisian electrical engineer Gustave Trouvé demonstrated a tethered model of a battery-powered electric helicopter at a meeting of the French Association for the Advancement of Sciences in Toulouse. The word "helico-pter" is constructed by two Ancient Greek words: ἕλιξ (hélix or "spiral") or ἑλικο (héliko or "twisted", "spiral-shaped", "bent") and πτερόν (pterón or "wing"). Therefore, the word helico-pter has two parts with logical meanings "helicon" for the spiral-shaped propeller and "pter" for the wings to fly.

But the new words like "quadrocopter", "multicopter" and "robocopter" have forgotten its etymological history of origin. The word "robo-co-pter" may be interpreted as a composition of "robot" and "heli-co" (the "co" makes no sense) and "pter". A correct linguistic composition would be "robopter", which sounds nowadays funny, so we will use "robocopter" in this paper.

3 What Is a Robocopter?

A robocopter is a flying robot. To get an impression of a robocopter, imagine a typical robot model from science fiction literature and film, like the famous R2-D2 created by George Lukas for the Star Wars movie in 1977. Put four arms with rotors onto the body of R2-D2, so it has rotors like a quadcopter, add some sensors and controllers for flight stability, additional communication interfaces, hardware, computers and software and then the R2-D2 can lift off and fly. Robocopters are quadcopters, multicopters and drones equipped with more intelligence (adaptability), more tools (than a typical quadcopter drone with a camera) and a higher degree of autonomy. Robocopters are not as dependent on a pilot for control, they can be used like an autonomous, self-driving car. A typical example of a command for a ride in a level-5 self-driving car (robot taxi) is: "Take me to my home." In level-5 autonomous cars no human intervention is necessary to get from A to B; even a steering wheel is optional—you get into the car and the robot taxi will take you to your home safely without a driver. A robocopter is very similar but with the extension of travelling in three dimensions. The robocopter is a robot aircraft, intended to work primarily with tools. Typical examples of a robocopter's services are: "Fly to the mailbox outside and bring me the newspaper.", "Bring me the table-tennis ball which is in the rain pipe on the roof", "I have forgotten my smartphone at the swimming pool, put it on the battery-charger in my home office".

Such services need many additional systems implemented in the robocopter, which go far beyond a typical quadcopter or multicopter used as a camera drone in the mass market since about 2006. The most important sub-system of a robocopter is a vision system with object recognition. Quadcopters nowadays already have a navigation system on board, enabling them to return to the lift-off platform very precisely, even with variable environmental conditions. They navigate using a GPS navigation system complemented with an image of the lift-off platform to easily compute the "return-to-home" position and the quadcopter is therefore able to land exactly on the lift-off spot. A robocopter needs to know more about its environment, e.g. it requires the coordinates to be mapped (coordinates of the mailbox at garden door, rain pipe on the roof, own smartphone, own swimming pool, inhouse home office, inhouse location of the battery-charger). But even if this mapping was pre-programmed and the robocopter is able to fly without crashing to these locations it cannot, as a robocopter, execute any actions, because it has no arms, fingers, box to carry things, etc. and furthermore would have absolutely no idea how a mailbox has to be opened, what differentiates a newspaper, which section of the roof is a rain pipe, how to search and grasp a table-tennis ball softly without destroying it, what is a swimming pool, a smartphone and a battery-charger and how to transport a smartphone safely and plug it to the charger. This sounds, in a way, that a robocopter will never be able to fulfil such tasks, because this robocopter must have such a lot of electronic, mechanical devices and batteries on board that the aircraft's gross weight is much too high for departure. However, this is not the case. Robocopters will become reality earlier than most people think.

4 Helicopter and Quadcopter

The quadcopter is based on the aeronautics and technology of the helicopter. The helicopter industry started in the late 1930ies and got a boost with the turbine jet engine in the 1950ies. Helicopters have the main advantage of vertical take-off and landing on a relatively uneven surface ground. They are used for many different services, e.g. rescue, flying ambulance, surveillance, military, police, security, transport, taxi, sports, construction, etc. The biggest helicopter today (the Mil-26) can carry a payload up to 20.000 kg with a gross take-off weight of 56.000 kg and a range of 800 km. Helicopters typically fly at a height between 800 and 1500 m above ground level with an average horizontal speed of 200–300 km/h. The cost to operate a 2-seat, 2-blade Robinson R22 helicopter is a minimum of US$ 200 per hour. The helicopter with greatest number of commercial sales is the 4-seat Robinson R44 Raven II with a cruise speed of 200 km/h, a maximum speed of 260 km/h, a range of 560 km, payload 340 kg (pilot and 3 persons), price about EUR 350.000, operation costs of EUR 180–280 per hour. For a private person is flying a helicopter quite an expensive hobby.

5 Quadcopter—The Self-flying Camera Drone

There is now an alternative for the hobbyist—a high-tech toy named "Quadcopter". For 50 years the main toy for boys and their fathers has been the train set. Generations have lovingly extended their train sets with beautiful and realistic miniaturised hand-made landscapes. But with the rise of quadcopters in about 2006, camera drone technologies, maker-scenes, cheap electronics, drone racing, etc. have turned the new generations from train sets to quadcopters. The Chinese company DJI, which was founded in 2006 by Frank Wang has today more than 50% of the world market in camera drones. Chinadaily.com.cn wrote on 13th of December 2016: "In its report published in May, PricewaterhouseCoopers estimated the global market for commercial applications of the drone technology will balloon from the current $2 billion to as much as $127 billion by 2020." In the article from the same source on 11th of January 2018 under the headline "China has transformed into a digital giant" Kai-Fu Lee and Jonathan Woetzel wrote, "As Chinese enterprises become increasingly technically capable, the country's market advantage is turning into a data advantage-critical to support the development of AI. The Chinese company Face++ recently raised $460 million, the largest amount ever for an AI company. DJI (a $14 billion consumer drone company), iFlyTek (a $14 billion voice recognition company) and Hikvision (a $50 billion video-surveillance company) are the world's most valuable firms in their respective domains." The global investments in artificial intelligence (AI), vision technologies (pattern, object and face recognition), voice recognition and flying devices are higher than ever before. Quadcopter camera drones brought into the market in the last 3 years offer incredible features, which include: HDR (high dynamic range) photo and video resolution with up to 1 inch and 32 megapixel camera sensor for 4K 60P video resolution, more flight autonomy with vision sensors for obstacle sensing to avoid crashes, gimbal technology which means that the camera sensor uses a special three-axis-electro-mechanical technology to ensure that the video recordings are smoothly stabilised and without distracting vibrations and shakes, ranges up to 7 km (the distance between pilot and quadcopter), flight time up to approx. 30 min, live high-resolution photo and video transmission to the pilot for a smartphone, tablet or a head mounted display (first-person-view video glasses or virtual reality (VR) goggles), horizontal flight speeds of 80 km/h (the record of racing drones is 262 km/h), software for flight control, planning and special effects, e.g. gesture control, to launch the quadcopter from hand palm and land there, to fly a preprogramed trajectory (e.g. rocket mode on DJI Spark: the quadcopter records a video of the pilots on the ground during a fast lift-off like a rocket; or asteroid mode on DJI Mavic Air: the quadcopter records a video similar to the rocket mode, but then the vision software morphs the ground into a sphere, so that it looks like the earth from space).

6 Drone Laws for Unmanned Aircraft Systems in Austria

In Austria, "Austro Control" (Austro Control Österreichische Gesellschaft für Zivilluftfahrt mit beschränkter Haftung) is the governmental agency for aviation law and it has regulated unmanned aircraft systems (drones) since 2014. For practical use a hobbyist drone pilot must obey the following rules for class 1 A drones (line-of-sight flight and drone under 5 kg):

(1) no threat or danger for persons or things;
(2) to fly a drone is allowed only in airspace class G using daylight after and before the civil twilight;
(3) maximum flight height is 150 m;
(4) maximum horizontal distance with direct sight between drone and pilot is 500 m;
(5) if the unmanned aircraft vehicle (drone) has a camera on board to record and store photos and videos on a chip, the drone pilot needs a license from Austro Control, which is given after a successful exam about aviation law and fundamental aviation procedures,
(6) if the drone is just streaming and not recording and flying in unpopulated rural areas, the pilot does not need a drone license,
(7) Austrian law allows "freedom of panorama", which means, that under copyright law everyone can take photographs and videos of objects, art work, buildings, etc. located in a public place, provided there is no copyright owner and a drone pilot with a license is therefore allowed to record videos of landscapes, even including people provided the people are they are sufficiently far away that they cannot be identified;
(8) for making photographs and videos from a drone you need a special drone liability insurance, which will cover the damage you can cause with a drone;
(9) a drone has always a subordinated status to all other aircraft types, e.g. if you see or hear a manned helicopter, you must land the drone;
(10) you can fly the drone in a wind speed of up to 36 km/h;
(11) quadcopters without a flight redundancy system are not allowed to fly over built-up areas. A hexacopter or octocopter by virtue of their construction have a flight redundancy system by an automatism for secure flight operation, which means that if one rotor fails, the drone won't crash;
(12) in the future every launch and landing will have to be documented in an aviation log book or in the App "Drone Space" by Austro Control;
(13) if you want to fly drones of class 1 C or D (class 1 means with direct sight between drone and pilot, C means using drones up to 5 kg in densely populated areas and D means using drones up to 150 kg in densely populated areas) you are required to pass a practical examination to demonstrate your skills, consisting of

(a) check of the drone before operation,
(b) to launch, to lift-off and go up,
(c) to hover with a 360° turn over the vertical axis to the left and to the right,
(d) climb flight,
(e) fast descent,
(f) climb flight to dedicated height,
(g) rectangle flight,
(h) fast flight and brake immediately in front of an obstacle,
(i) flying around an obstacle,
(j) climb flight to maximal height of the drone,
(k) flight to the maximum 150 m height to maximum range as long as the drone is in sight,
(l) simulated crash of the transmitter—automatic landing at the home position,
(m) launching with a predefined flight program,
(n) landing manoeuvre.

Source: www.austrocontrol.at: „Anlage P zu LBTH 67: Flugprogramm für den Praxisnachweis der Kategorie C und D von Klasse 1 uLFZ Piloten, 14.7.2015". The courses for class 1 A offered in Austria last 2 days and end with a 60 min multiple choice exam processed by Austro Control.

7 From the Flying Camera to the Flying Robot Drone

The flying camera drone has another aspect of fascination. It realises an old dream of mankind—to rise into the air and fly as described in the ancient Greek myth of Ikaros with his father Daidalos. This old myth was popular in the era of Homer. The Roman poet Ovid (43 v. Chr. –17 n. Chr.) and later the German priest and author Gustav Schwab (1792–1850) brought this great story into our modern collective memory. Daidalos and his little son Ikaros have been prisoned by the King Minos in the labyrinth of Minotaurus at the Greek island Crete. Daidalos was a brilliant inventor and constructed wings for him and his son. He used a frame for the wings and attached feathers with wax to the rack frame. Daidalos warned Ikaros that he should not fly too high because the sun could melt down wax and not too low because the water could touch the feathers, which would cause a crash in both cases. But Ikaros enjoyed flying like a bird and was so fascinated he became careless. Ikaros came too near to the sun, the wax melted and Ikaros died by crashing into the sea. With camera drones and first-person-view (FPV) glasses it is